W9-BSN-813

BOSTON
ON
FIRE

BOSTON
ON
FIRE

A HISTORY OF FIRES AND FIREFIGHTING IN BOSTON

STEPHANIE SCHOROW

COMMONWEALTH EDITIONS
Beverly, Massachusetts

Copyright © 2003 by Stephanie Schorow
All rights reserved. No part of this book may be reproduced in any form or by any
electronic or mechanical means without permission in writing from the publisher,
except by a reviewer who may quote brief passages in a review.

Second paperback printing, January 2007

ISBN 13: 978-1-933212-01-2
ISBN 10: 1-933212-01-2

Library of Congress Cataloging-in-Publication Data
Schorow, Stephanie.
   Boston on fire : a history of fires and firefighting in Boston / Stephanie
Schorow.
   p. cm.
   ISBN 1-933212-01-2
   1. Fire extinction—Massachusetts—Boston—History. 2. Fires—
Massachusetts—Boston—History. I. Title.
   TH9505.B7S34 2003
   363.37'09744'61—dc21   2003012612

Cover and interior design by Kimberly Glyder.
Images in jacket montage: Top, Chelsea in flames, from the collection of
Herbert C. Fothergill, courtesy of Mr. Fothergill; lower right, rescue from the
Cocoanut Grove, from the collection of Bill Noonan, courtesy of Mr. Noonan,
the Boston Public Library, and the *Boston Herald*; lower left, firefighters at the
Vendome Hotel fire, Kevin Cole photo, courtesy of the *Boston Herald*. Image on
back of jacket: "Boston in Flames," Currier & Ives lithograph of the Great Fire
of 1872, courtesy of The Bostonian Society/Old State House.

For other illustration credits, please see page 232.
Printed in United States of America.

Published by Commonwealth Editions,
an imprint of Memoirs Unlimited, Inc.,
266 Cabot Street, Beverly, Massachusetts 01915.

Visit our Web site: www.commonwealtheditions.com.

CONTENTS

FOREWORD   *Leo Stapleton*   vii

INTRODUCTION   *The Fire Trail*   x

CHAPTER 1   *Built to Burn: Kindling Fire in Early Boston*   1

CHAPTER 2   *Fires of Wrath: The 1834 Convent Fire and the
1837 Broad Street Riot*   13

CHAPTER 3   *Built in Boston: The Soul of the Old Masheen*   30

CHAPTER 4   *Strike the Alarm: The Nerves of Boston's
Fire Alarm System*   48

CHAPTER 5   *The Great Fire of 1872: A Disaster Foretold*   67

CHAPTER 6   *Twice Burned in Chelsea: The 1908 and
1973 Conflagrations*   103

CHAPTER 7   *The Cocoanut Grove: Heat, Smoke, and Panic*   126

CHAPTER 8   *Death in the Vendome: The Perils of Firefighting*   176

CHAPTER 9   *Mr. Flare and the Ring of Fire: Arson in Boston*   191

AFTERWORD   212

REFERENCES   219

INDEX   233

ACKNOWLEDGMENTS   242

A fire in Boston, 1832

FOREWORD

During the nearly forty years I served with the Boston Fire Department, I found that most of the department's rich history, including the history of the many Boston fires, was based on rumors, innuendo, and occasionally outright falsehoods. Stephanie Schorow's marvelous historical account of Boston fires and fire fighting has made me feel quite sheepish that I didn't take enough interest in those who had preceded my colleagues and me, and in how the organization has evolved into one of the most respected in the entire fire service.

Covering a wide span of history—starting with the first primitive attempts to control fires, shortly after the Puritans made their way to Boston, and continuing into the present—is certainly an awesome task for any author. It involves dedication, determination, and a love of the subject matter, along with a style of writing that captures the reader's attention.

There have been thousands upon thousands of serious fires in Boston and the surrounding communities through the centuries, and Ms. Schorow has selected a number of the most serious and prominent incidents, often involving loss of lives and tremendous property damage. These disasters represent the difficulties and dangers firefighters frequently encounter in the course of their duties.

A great number of firsts are associated with the Boston Fire Department, including the purchase of the initial vehicle designed for fighting fires, back in the 1600s; the first telegraph fire alarm system in the world, in 1851; and many other innovations related to respiratory protection and other personal protective equipment for firefighters.

The representative number of fires Ms. Schorow chose to research in depth is a very wise selection. Until I read this book, I only vaguely remembered hearing about the 1834 Ursuline Convent fire in the Charlestown District. The fascinating circumstances leading up to the incident made me realize just how much prejudice, both ethnic and religious, existed in those long-ago days—not unlike what still remains today in many parts of the country and the world. The Broad Street Riot, just a few years later, which is so well described, further underscores this view.

Ms. Schorow describes in detail the Great Fire of 1872, which destroyed much of Boston's financial and commercial district; the 1942 Cocoanut Grove disaster, which led to the largest loss of life in a nightclub fire; the 1908 and 1973 conflagrations in the nearby city of Chelsea; and the 1972 fire at the Vendome Hotel, in which nine firefighters died when the hotel partially collapsed. For her discussion of twentieth-century fires, wherever possible Ms. Schorow has personally interviewed survivors as well as fire personnel who participated in rescue and fire-fighting efforts.

A massive number of arson fires plagued Boston during the early 1980s, when a severe budget crisis led to the termination of many firefighters. Ms. Schorow has managed to interview some of those behind the events of this bizarre period—none of whom were laid-off members of the fire department.

The afterword briefly describes other notable, serious incidents, including the Bellflower Street fire, the Prudential Tower fire, and major fires in surrounding communities. Ms. Schorow's personal reflections about how her research has affected her own views of the fire service are quite poignant.

Reading the names of so many figures, prominent in the history of the country, who lived and worked in Boston, and their relationship to the fire service, makes one realize how large a part this city played, from a historical point of view, in the establishment of the United States of America, as well as in the evolution of fire fighting. The details of fire fighting operations, as well as many personal reflections of those involved, make for a book that anyone with a love of the fire service should cherish. It should also be most instructive to the general population, who know so little about what firefighters actually accomplish on their behalf.

LEO D. STAPLETON
Fire Commissioner
Chief of Department
(retired)
Boston Fire Department

FIRE: AN INTRODUCTION

Almost every visitor to Boston runs across the city's famous Freedom Trail. This path, marked by a red line on the pavement, winds through the city, linking the points where history was made: Paul Revere's home, the Old North Church, the Old South Church, Bunker Hill. But there's another trail in Boston, unmarked and unnoticed, that weaves through the narrow streets on a more somber journey. It is Boston's fire trail, a path of points where another kind of history was made—the history of fires, firefighting, firefighters, and fire lore. Like the Freedom Trail, the fire trail starts in Boston and continues in spirit through the nation. Not only did fire change Boston's geography, laws, and the lives of so many of its people, but the effects of those changes also rippled through the continent. Perhaps only those who daily face the threat of fire fully understand the impact of Boston's fires on firefighting everywhere.

"Fire fiend." "Niagara of destruction." "The giant." "The monster." "Possessed of an evil spirit." These are among the names that early Bostonians gave to the destructive force that returned, again and again, to consume their homes and businesses. When frequent fires ravaged the young town, Boston's Puritans declared that the flames were the work of a wrathful God, a reminder of the fickle and fleeting nature of earthly existence. God's wrath notwithstanding, Bostonians faced down the fire fiend with building regulations, innovations in fire apparatus, and volunteers who began a tradition of self-sacrifice for the public good.

Boston has its share of fire "firsts": the first paid fire department, the first building codes, and the first municipal fire alarm system. All these advances made the city justifiably proud of its efforts in keeping the fire fiend at bay. But hubris can end on a fiery bier. In 1872 a burgeoning Boston, fat and complacent in the post–Civil War boom, ignored warnings that the city was growing too fast, too soon. The result was the Great Fire of 1872, a conflagration that left the commercial district in ashes and reshaped the city's downtown. Few residents, strolling today amid the bustle of the shops at the city's Downtown Crossing or jabbering on cell phones as they scurry into large office buildings, realize the full devastation of the fire that torched the ground under their feet. It was a inferno so fierce that without the Atlantic Ocean as a fire break, it would have, as modern firefighters like to joke, burned Berlin.

Boston also seems to have had more than its share of fiery tragedies. In November 1942, in less than fifteen minutes, fire swept through the popular Cocoanut Grove nightclub, killing or maiming nearly 500 people and changing the lives of thousands of others. Chapter 7 breaks new ground in detailing theories on the mysterious cause of this fast-moving blaze and tracking the fate of those who survived its horror. The fire made medical history and changed fire codes throughout the country, as officials vowed to end such tragedies for all time. And yet, in a troubling echo of the conditions that produced the Cocoanut Grove fire, a terrible fire swept through a Rhode Island nightclub in February 2003, killing about 100 people and injuring dozens more. The similarities between the two infernos are both eerie and tragic.

In 1972 the city mourned nine firefighters killed when the century-old Vendome building, weakened by a fire, suddenly collapsed, underscoring another hazard of fire that has brought death since the very early days of the city. In the early 1980s, the city earned the dubious title "arson capital of the nation" when a string of bizarre arson fires terrorized the city.

But even in pre-Revolutionary times, when Boston suffered more fires than any other American settlement, daring men with limited equipment and virtually no personal safety gear ran into burning buildings, scaled flaming walls, and pulled victims from roaring furnaces. On September 11, 2001, the nation was riveted by the terrible losses suffered by firefighters attempting rescues in the World Trade Center in New York City. Such courage and disregard for personal safety date back centuries to a time when volunteer firefighters were seen as embodying the very spirit of America.

One such firefighter was nineteenth-century District Chief John Francis Egan—a man who, newspapers said, "seemed to delight in danger." In March 1893 he thrilled the city with a daring escape from the top of a blazing building. While fighting a fire at the company of Brown, Durrel and Co., Egan had climbed through a skylight onto the roof to turn on the roof hydrant when flames burst through the skylight, cutting off his escape route. As a horrified crowd watched, he grabbed a thick electric cable that stretched across Kingston Street to the Holmes building on the opposite side. He swung off the roof and, hanging by his fingers, attempted to cross to the other building. He reached the midway point when the sag of the cable stopped his process and he hung, eighty feet above ground, as people below scrambled to create a net. Meanwhile, three men on the Holmes building attached a rope to the cable and began to lower it. Egan used the pitch to slide downward, but the rope ran out twenty feet from the ground. His strength failing, he dropped. The net broke his fall.

Five years later, on February 5, 1898, in the wee hours of the morning a fire was reported on the fourth floor of a five-story brick building occupied by George Bent's bedding factory on Merrimac Street. Handicapped by a recent snowfall, firefighters frantically worked to get water flowing from the frozen hydrants. Members of Engine Company No. 7, including Egan, ran to the fourth floor to operate hose lines. The fire was just about under control when the roof collapsed into the fifth floor and all the floors came down. Men started to frantically dig into the debris and the snow. The bodies of six firemen were pulled out; among the last was that of Egan.

Despite its perils, firefighting in Boston—as in so many other areas of the country—has became a tradition handed down from father to son and more recently from father to daughter. In many Boston firehouses, the children and grandchildren of firefighters choose to join the ranks, fully aware of the good they can do and the dangers they face. That Boston's fire department clings proudly—and stubbornly—to tradition is revealed by one little fact: There is no Engine Company No. 1 in Boston. Nor is there an Engine 6, 11, 12, 13, 15, or 19.* Yet the thirty-three engine companies are numbered up to 56. That's because companies tenaciously clung to "their" numbers, even when companies were disbanded or added. Engine 7 remains the city's oldest engine company.

The fires of Boston are never far below the surface of modern life. Like the Freedom Trail, the fire trail is one of actual places and could be walked in a day. A visitor might begin on Washington Street, the place of several colonial-era conflagrations and the location of the Old South Church, a survivor of the 1872 blaze. Just blocks away is fire alarm box number 52, where that huge fire was first reported. If you continue across the Fort Point Channel, you reach Boston's Fire Museum, housed in a former fire station that was built on landfill from 1872 rubble. From there it's less than a twenty-minute walk to Bay Village, the site of the terrible blaze inside the Cocoanut Grove nightclub. Another fifteen minutes takes you to into Boston's Back Bay and the Commonwealth Avenue mall, with its monument to the fallen firefighters of the Vendome Hotel fire. The rebuilt Vendome itself is just across the street. As in so many other aspects of Boston lore, the city's fire history remains very much in view—if you know where to look.

In 1666, the Puritan poet Anne Bradstreet wrote an elegy for her home, which had been destroyed by fire: "That fearful sound of 'Fire' and 'Fire'/Let no man know is my desire." She would not see that dream fulfilled, nor would the people of Boston in the centuries that followed—not even today. The story of Boston's firefighting is one of ingenuity, tenacity, tragedy, and courage; the history of Boston's fires tells the history of the city itself. This book is a journey along the fire trail told through the eyes of those who fought, witnessed, or survived the fires of Boston.

---

*Engine companies numbers 23, 25, 26, 27, 31, 35, 36, 38, 40, and 43–47 have been disbanded.

1

BUILT TO BURN

Kindling Fire in Early Boston

A cold, dry wind blew through Boston early in the evening of October 2, 1711. Mary Morse huddled near a fire in her ramshackle home on Cornhill Row, near the center of town, and tried to keep her fingers moving. The town, settled in the 1630s as the Puritan vision of "a city on a hill," was now a bustling commercial center of about 10,000 citizens, crowded into a narrow peninsula. Ships from France, Spain, and Portugal brought in cargoes of wine and brandy, silk, laces, and salt to large wharves built into the cold Atlantic; ships departed with loads of cod. Immigrants from Scotland and Ireland were starting to add to the numbers of the original English settlers, and the slave trade was bringing more black servants. Stern Puritan leaders, such as the Reverend Increase Mather and his son, Cotton, also a prominent clergyman and leader, still held sway over the governance of the town, although people of other religions were drifting in. Visitors commented on the town's prosperity, but the houses, made of thin, small cedar shingles nailed against frames and filled in with bricks, were built nearly on top of each other. As visitor Jasper Dankers noted, Boston was liable to fire "as already happened several times; and the wonder to me is that the whole city has not been burnt down, so light and dry are the materials." Indeed, since 1653 the young town had suffered seven "great fires."

The commercial expansion had, moreover, widened the gap between rich and poor, and Mary Morse, described as a poor Scottish woman, had to "pick" oakum to bring in a few pennies. She would pull apart, fiber by fiber, old pieces of hemp rope from used ship rigging. The loose fibers would be twisted back together and used in packing or coated with coal tar and used for caulking. Picking oakum was hard, tedious work, and so what if she had a wee drink or two . . . or three? She didn't notice at first that sparks from the fire had landed on the fluffy piles of picked oakum. When she did, she cursed and tried to stomp out the flickering flames. But the dry oakum burned quickly, and in a few minutes she had no choice but to flee the tenement, screaming, "Fire!"

Others took up the cry. As the word spread, the church bells began to chime, alerting the citizens that a fire had been spotted. Responding to a fire was a civic duty for all, even if some people already had been designated as "fire men" or "fire wards." Every household was required to keep leather buckets on hand; often the buckets were marked by the family name or coat of arms and carefully collected after a fire. If a bucket brigade couldn't keep a fire from spreading, people would pull houses down with large hooks, hoping to create a fire break. By 1711 the town also had a few "injines," primitive contraptions that could produce a stream of water when pumped by hand. Now the fire wards were hastily trying to get the machines to Cornhill Row, while others started to form bucket brigades.

It was already too late. A drought had left the city tinder dry, and the fire spread from building to building down Cornhill (roughly where Washington Street in downtown Boston is today) as terrified residents attempted to grab whatever possessions they could and flee. "The Flames took Hold of the Neighboring Houses . . . notwithstanding all application and diligence to extinguish and prevent the spreading thereof by throwing of Water and Blowing up of Houses," the *Boston News-Letter* of October 1–8, 1711, later reported. The flames roared to School Street, Dock Square, and the Pudding Lane (now Devonshire

Street), cutting a swath of destruction through one of the most thickly settled and wealthy parts of town.

The flames reached the Old Meeting House, built in 1630, the year the town was founded. Several sailors climbed into the steeple to save its precious bell. They struggled to cut the ropes as the flames grew closer, but before they could climb to safety, fire cut off all means of escape and they fell into the inferno. Only their blackened bones were recovered. Nearby was the venerable Boston Town House, the seat of government and a marketplace—the very center of the city's commercial and intellectual life. Here the first colonists defied the English governor by demanding self-rule; here the Puritans finally relented and allowed in the first Quaker church. After fifty years, the building came to its end in a fiery blaze. Some brave souls managed to save portraits of the queen that were inside, but the building was lost.

As the fire continued to spread, at least eight people were killed and many wounded by falling timber. "Many poor men . . . were killed by the blowing up of Houses; or by Venturing too far into the Fire, for the Rescue of what its fierce Jaws was ready to Prey upon," Cotton Mather lamented in his sermon after the fire. A total of 100 buildings were destroyed, leaving 110 families homeless. The flames could be seen for a distance of twenty leagues, yet somehow, by 2 A.M. the next day, the fire was put out. Not only did people lose homes, but thieves made off with many of the possessions that had been moved into the streets before the advancing fire.

With grim resolution Cotton's aging father, Increase Mather, stepped up to the pulpit of the North Meeting House the next Sunday. With a fire-and-brimstone flourish, he thundered, "Desolating Fires are an awful judgment, but the Lord's Kindling."

History does not record what happened to poor Mary Morse, but she was denounced from the pulpit. "The fire we now Bewail this day, is supposed to be occasioned by a wicked, drunken Woman. And let those who have their Houses taken from them, be thankful that it was not done sooner." Mather was just warming up:

Has not God's Holy Day been profaned in New England?
Has it not been so in Boston this last summer, more than ever
since there was a Christian here? Have not burdens been car-
ried through the streets on the Sabbath Day? Have not bakers,
carpenters and other tradesmen been employed in servile works,
on the Sabbath Day? When I saw this . . . my heart said, Will
not the Lord for this kindle a fire in Boston?

Somehow, when the ashes cooled, Bostonians—sinners and sinless alike—picked up and moved on. They used the rubble to fill up Long Wharf, they rebuilt homes and shops with bricks, and following a tradition established by the very first fire, they passed more regulations. They had been through huge fires before and would endure still more before the century's end. For, as fire historian Paul Ditzel would say 250 years later, "Boston was built to burn." While other cities in the American colonies also endured ferocious blazes, Boston was targeted by fire more than any of them.

The first fire of record occurred barely eight months after the town was first settled. "About noon the chimney of Mr. Thomas Sharp's house in Boston took fire, the splinters not clayed at the top and taking the thatch, burnt it down," Massachusetts Governor John Winthrop noted in his diary. Afterward, "we have ordered that no man shall build his chimney with wood, nor cover his house with thatch, which was readily assented onto," Lieutenant Governor Thomas Dudley wrote. Chimneys then were made of splints of wood, held together and covered with mud and clay. If the clay flaked off, however, the chimney could ignite.

For seventeenth- and eighteenth-century Bostonians, an open flame was a constant in their lives. Fire was the only source of heat and light; its heat forged metals, baked bread, and turned clay into useful vessels. They burned forests to clear the land for planting life-sustaining crops. The first settlers had come to this country with the intention of building heaven on earth; the abundance of wood in New England, in contrast to fuel-starved England, further confirmed that the Puritans were doing the Lord's work. "Here is good living for those that love good fires," said observer Francis Higginson, who boasted that wood was so plentiful that a "poor servant" here could easily obtain as much fuel as a nobleman in England. But by 1637 wood was at a premium in Boston, owing to the town's location on a peninsula far from any forests. Moreover, "good fires" could quickly get out of hand. During a particularly cold winter, Cotton Mather wrote in his diary that his ink had frozen in the stand near the stove: "My Ink in my very pen suffers a congelation: but my wit much more."

On a bitter cold January day in 1653, a fire broke out near the town's waterfront and spread through the wooden warehouses to King Street. Not only did the blaze jump from house to house, but sparks ignited barrels of stored gunpowder inside. "It was a wonderful favor of God the whole town was not consumed," one observer noted. By the time the fire burned itself out, three sleeping children had died and about one-third of the citizenry were homeless. Officials promptly passed an elaborate list of rules that affected almost every citizen. They included:

Every house should have a ladder that reaches to the ridge of the house; otherwise, owners faced a penalty of 6 shilling 8 pence.
Every household should have a twelve-foot or longer pole with "a good large swob [swab] at the end of it" to be used to snuff sparks on the roofs.
The town meeting house would be provided with "good strong Iron hooks" plus chains and ropes, which would be hung at the side of the building to be ready "in case of fier." Also, selectmen shall provide "six good and long ladders" for the town's use.
No house could be pulled down to create a fire break without the consent of the town leaders, but house owners could not seek compensation, either.

The rules underscored the two great "givens" in the Puritan mind: the laws of God and the rules of man. "If fires could be legislated out of existence, then Boston would have been spared," Paul Ditzel declared. Boston already had another legal oddity: the country's first antismoking rules. In 1638 Boston selectmen outlawed smoking *outside* because "fires have been so often occasioned by taking tobacco" out of doors; it was better, they thought, to restrict smoking to inside homes. Another significant law was passed in October 1679, when the Massachusetts court ordered that all new dwellings be constructed of stone or brick and covered with slate or tile, or else the builder would forfeit double the value of the buildings. But brick was expensive and many could not afford the extra burden, so the rules were suspended and modified. Wooden houses continued to be built. The law was not really enforced until 1683. By then, however, Boston had suffered two more major fires.

In November 1676 a fire was apparently caused when a tailor's apprentice left a candle burning. It fell over and ignited the house of Mr. Wakefield near the Red Lion Tavern, and flames spread through the area surrounded by Richmond, Hanover, and Clark Streets. About fifty houses and other structures were destroyed, including Increase Mather's own church, the North Meeting House. (It was rebuilt the following year.) Years later Cotton Mather insisted that his father had a premonition of the disaster when he preached on the previous Sunday that a fire was coming: "Oh! Lord God, I have told this people that thou art to cut off their dwellings, but they will not believe." The city also suffered a string of arson fires in 1679, set by what Boston fire historian Arthur W. Brayley called "determined fire bugs."

Perhaps Increase Mather was onto something. Despite all its precautions, Boston suffered at least one major fire a year—more than all other American cities combined. Even minor fires have had profound impact on the development of the city. The State House (built to replace the Town House destroyed in 1711) was itself destroyed by fire in 1747. Rebuilt, it suffered another bad fire in 1832. Faneuil Hall, the famous market and meeting place designed by Peter Faneuil, burned in 1761 and was rebuilt in 1763. "At the rate Boston burned, it was a miracle that anything remained to be designated a historical monument," Paul Ditzel observed.

*Seventeenth-century
fire in Boston*

The greatest weapon against fire, however, was not rules or machines; it was the organization of what became a peculiarly American institution, volunteer firefighting. Boston, New York City, and Philadelphia have each—with merit—claimed the honor of producing the first volunteer fire department. That honor, however, depends on one's interpretation of a variety of supposed "firsts."

After the 1676 fire, Bostonians decided to invest in new technology ordered from London: a wooden box, about three feet long and eighteen inches wide, with front and rear handles that let it be carried to fires. Filled with water by a bucket brigade, an interior pump, operated by hand, shot a stream of water out of a flexible hose. The town also realized that the new machine needed someone with the skill to operate it, so Boston selectmen

"Hand tub"
fire engine

passed a resolution on January 27, 1678, designating Thomas Atkins, a carpenter, to be foreman of "ye engine lately come from England" and to choose assistants whom the town would pay "for their pains about the work." Twelve men were thus named; they were, in effect, the country's first paid firefighters. One of them was even compensated by the town for his injuries on the job: Richard Whieraye was unable to perform heavy physical work after being pinned down by massive beams during a fire. He was awarded £20 and a license to keep a pub. Boston now enjoyed the best in seventeenth-century firefighting technology, twenty years before Paris bought its first fire engine.

Successors to Atkins were also designated in 1683, 1686, and 1703. By 1707 two other companies were formed to handle two additional engines. In the wake of the terrible 1711 fire, citizens were influenced by the organized fire response of systems in Amsterdam, which put men "under the same discipline as soldiers" and advised that strangers venturing into a street where a fire has broken out "immediately be clipt into Gaol." Ten fire wards were appointed and given authority to organize firefighting efforts, arrest looters, and order citizens to pitch in. By 1720 Boston had the beginnings of a modern fire department, with six engines and twenty firefighters.

Another kind of institution was established. In September of 1718, citizens organized the "Boston Fire Society," the country's first mutual aid organization. Members of the society pledged to help one another out "in case it should please Almighty God to permit the breaking out of fire in Boston where we live." That meant they would rush to fight fires at members' homes and, just as important, guard against looting. Each of the twenty members pledged to bring two buckets, two large bags, and a bed key to every fire. (Beds were then one of the more valuable of colonial furnishings and a bed key assisted in their rapid disassembling during a fire.) Because members went to the aid only of other members, this could not be considered a municipal fire company, but it had a significant impact on one imaginative Bostonian.

Young Benjamin Franklin was fearfully impressed by the chaos of the 1711 fire. Born in Boston on January 17, 1706, he later moved to Philadelphia, where he began publishing a newspaper in 1733 and expounding on fire prevention. He organized the Union Fire Company in 1736, in which, like the Boston Fire Society, members volunteered to rush to the aid of other members. Franklin, never shy about the spotlight, was the society's first chief.

Boston's Atkins and Franklin's society notwithstanding, Peter Stuyvesant, the pugnacious Dutch governor of New Amsterdam—later Manhattan—gets the official credit for establishing the first volunteer fire unit. At least his somber visage looks out proudly from the U.S. postage stamp issued in 1948 that celebrated the "300th anniversary of volunteer firemen." In 1648 Stuyvesant appointed four property owners as fire wardens to enforce some of the community's fire prevention restrictions on buildings and chimneys. About ten years later he established the "Rattle Watch," who would patrol the streets at night and sound noisemakers in case of fire. But because volunteer companies weren't organized to fight fire until New York got its first hand-pumped fire machines in the early 1730s, Stuyvesant could more correctly be credited with creating the first fire prevention officers.

By 1760 Boston had nine fire companies, each with a captain and eleven to eighteen men. They would be put to their greatest

test in colonial Boston's last great fire—and probably its worst—
a fire that would plant the seeds for the American Revolution.

In March 1760 a cry of "Fire" rang out at about 2 A.M. when
flames were seen in the Brazen Head, a large house and tavern on
Cornhill, owned by Widow Mary Jackson and her son William.
Within ten minutes the first fire company had pulled its rig to the
scene. The Jackson house was gone, but the men were confident
they could keep the fire from spreading. But the flames spread to
Pudding Lane, now Devonshire. Engine companies, already
tagged with names, ran to the scene, Old Prison Engine 7, Hero
Engine 6, and Old North Engine 3 among them. Members of the
various mutual aid societies were also arriving, preparing to rescue the property of members. At 4 A.M. Nathaniel Ames looked
out his window and "beheld a blaze big enough to terrify any
Heart of common Resolution, considering such valuable com-
bustibles fed it." The conflagration, described by another
observer as "a perfect torrent of fire," devoured shops on Pudding
Lane and homes along King Street and Quaker Lane (Congress
Street), cutting a swath of destruction eastward toward the
waterfront. Now visible fifty miles away, the conflagration licked
at the beams of Long Wharf and consumed ten ships docked
there.

"The distressed inhabitants of those buildings, now wrapped in
fire, scarce knew where to take refuge from the devouring
flames," Town Clerk William Cooper wrote in the *Boston Post Boy*
on March 24, 1760. Witness Samuel Savage, who watched the
flames from Fort Hill, wrote: "I can say without exaggerations
that I never in my life was in greater storm or snow nor knew it
snow faster than that fire [that] fell all around us." Neighboring
towns rushed in with men and equipment but, as Cooper said,
"The haughty flames triumphed over our engine—our art—and
our numbers." By the time the blaze had burned out, after ten
hours, it had consumed about 350 homes, shops, warehouses,
and other buildings. Hundreds of people, both rich and poor,
were homeless. Estimates of the loss range from £100,000 to
£300,000, a staggering amount for the town, and the worst fire to

date in the American colonies. "Not one life has been lost and few wounded," Cooper reported. But with the town treasury quickly emptied of its £3,000, Boston was in dire need. A petition for help was sent to King George III; it was acknowledged two years later with only the message that it had "been graciously received by his Majesty." Bostonians rebuilt their town, but residents seethed at the indifference of the English monarch, a resentment that would bloom into armed rebellion within fifteen years.

After the 1760 fire, the city was divided into fire districts, each overseen by a fire ward. (John Hancock and Samuel Adams were among the prominent Bostonians who acted as fire wards.) By 1763 sixteen fire wards maintained ten engines, each with a company of thirteen to twenty men. The system gained a reputation for quick response and efficient organization. Yet major fires continued to break out well into the next century. In 1787 a blaze soon raged out of control as a dry northeasterly wind blew sparks from shingled roof to shingled roof. Eighty-six families were burned out and a church destroyed, prompting an unknown poet to lament: "Once renown'd beloved City/How I deprecate thy

1797 engraving of a fire on State Street

Loss/How thy Glory Lies in Ashes/Soon thy Gold is turn'd to Dross!" But on the eve of the inventive nineteenth century, Bostonians would rely on technical prowess, not divine intervention, to control fires. In 1792 the Massachusetts Charitable Fire Society was organized not only to help fire victims but for "stimulating genius to useful discoveries tending to secure the lives and property of men from destruction by that element." This society, whose membership has included such luminaries as Samuel Adams, Charles Bulfinch, Josiah Quincy, and Paul Revere, would operate in some form for the next 200 years.

In 1822, two years after its population passed the 43,000 mark, Boston became a city, leaving behind its historic form of town meeting. But it remained a burning zone. As of 1825 Boston "seemed doomed to be destroyed by extensive and disastrous conflagrations," said David Dana, a contemporary Boston fire official and fire historian. City officials were also alarmed at the inefficiency of the bucket brigades organized by the fire wards. Mayor Josiah Quincy, elected in 1823, was particularly interested in modernizing the fire ward system, specifically by replacing bucket brigades with water drafted by hoses from municipal water sources. He had, as his son later wrote, "most determined opposition to encounter. The old ways had been good enough for their fathers, and why not for them?" Hoses were "denounced as absurd and almost wicked," while the modern engines in New York and Philadelphia were "an affront to the mechanics of Boston." Quincy's effort to establish cisterns at convenient points was considered a monument to his extravagance. When the fire ward captains resigned in protest, Quincy made his move; he abolished the fire ward system and established the first official citywide fire department in 1825. The first "chief engineer," Samuel D. Harris, was appointed to direct it in 1826.

Boston might have been built to burn, but the city and its firefighters would spend the next centuries defying that fate.

2

FIRES OF WRATH

The 1834 Convent Fire and the 1837 Broad Street Riot

Neither pleading nor promises nor copious tears could change her father's mind. Young Louisa Goddard would be sent to a Catholic boarding school, there to remain until she was 20! Mr. Goddard, who believed in the widest liberty for men, unceasingly lamented the growing independence of women. So, although a good Protestant, he decided the nuns would save him the trouble of educating his rambunctious daughter in the habits of strict submission to authority.

So at the end of July in 1834, a weeping Louisa said good-bye to her dolls and kitten* and was lifted into a carriage and driven four miles from her Dorchester house to a large, grand building on a hill, surrounded by fields and orchards, in what was then Charlestown. She little dreamed that her Catholic school days would be cut short in less than two weeks in an attack inflamed by ethnic and religious hatred—an attack in which fire was the weapon and firefighters were among the attackers.

Today, after hundreds of St. Patrick's Day parades, the political rise of the Kennedy clan, a succession of Irish mayors—not to mention the famed Celtics basketball team—a Boston hostile to Irish Catholics is almost unimaginable. But in the early part of the

---

*Whitney's narrative implies that she was 11 or 12 in 1834, but according to her birth date, she was 14 when she attended the convent school.

nineteenth century, Irish immigrants coming to Boston to escape the desperate poverty of their homeland found few to welcome them. Although the Boston Brahmin elite were perfectly willing to hire Irish maids and workmen, they regarded the newcomers with contempt. Working-class Bostonians feared that the desperately poor newcomers would take away jobs.

Boston's volunteer firefighters, in particular, seethed with resentment at the unskilled immigrants who would work harder for less money; they saw them as a threat to both their economic standing and prestige, which had begun to fade. Once considered the romantic embodiment of the American can-do spirit, the men who "ran with the machines" were increasingly regarded as belligerent toughs who cared more about getting to a fire first than putting it out quickly. Convenient targets for their frustrations were immigrants, who had even less social standing than they

Cartoon expressing anti-Irish sentiment from D. C. Johnston's Scraps 1835

had. Boston's mostly Protestant population then had deep-seated biases against Catholicism. Catholic religious practices seemed both exotic and threatening; local religious leaders often railed against "Romanism" and "popery" even as the number of New England Catholics increased.

For the Irish continued to arrive. Boston Irish Catholics numbered about 2,000 in 1820; by 1830, that population had grown to 7,000. Thousands more would come during the Potato Famine (1845–1848) and would continue to come for a few decades. While the Irish had yet to achieve political clout—only 200 Irish residents were registered to vote in Boston as of 1834—many Bostonians were loudly complaining about the insidious influence of the newcomers. One of the most vocal critics was the Reverend Lyman Beecher—ironically, the father of abolitionist Harriet Beecher Stowe, whose book *Uncle Tom's Cabin* called for the emancipation of African-American slaves. A leader of an orthodox, conservative movement then sweeping the region, Beecher railed against Catholic schools that taught Protestant students. All of Boston knew exactly whom Beecher was attacking. Just over the river, in the township of Charlestown, Ursuline nuns had established a prestigious school open to both Catholic girls and the daughters of wealthy Protestants.

The Ursuline sisters, a Catholic order devoted to education since its founding in the sixteenth century, had sent sisters (many of them Irish) from Quebec to Boston at the request of Boston's Catholic bishop, Benedict Joseph Fenwick, who was eager to provide education to the growing numbers of Catholic children. In 1824 the Ursulines established a boarding school in Charlestown on a twenty-four-acre plot called Ploughed Hill, which they renamed Mount Benedict in honor of the bishop. Officially "the Ursuline Community, Mount Benedict, Charlestown," the school was usually called "the nunnery" or "the convent" by locals. In a spacious three-story building, a dozen nuns and several laywomen taught reading, writing, arithmetic, geography, history, drawing, and needlework (plain and fancy) for the then-pricey tuition of $125 a year, plus $4 for ink, quills, and paper. In fact, the school was so expensive that most poor Irish families could not afford to send their daughters there. The Mother Superior, 41-year-old Sister Mary Edmond St. George (the former Mary Anne Moffatt) was, by all accounts, a strict, assertive, even arrogant woman, who nonetheless was held in high esteem by the bishop and who, in a more modern era, might have run her own business. St. George was "a woman of masculine appearance and character, high-tempered, resolute, defiant, with a stubborn, imperious will," according to Patrick Donahoe, founder of *The Pilot,* Boston's Catholic paper. Young Louisa was in awe of Mother Superior at first sight, saying later, in her account of her days at the school: "She carried herself with royal uprightness and dignity that compelled deference from all who approached her. Servants followed her with shawls, cushions, and parcels and their deferential manner made her grand air more apparent." John Buzzell, a local brickmaker who often bragged of beating up Irishmen, called the Mother Superior "the sauciest woman I ever heard talk."

In 1834 the Ursuline school was prospering. Most of the forty students who attended the school were Protestants from wealthy families (one was the daughter of a judge), who worshipped separately from the Catholic girls. Although deeply homesick, Louisa took great interest in the seemingly exotic lives of the nuns and started to make friends with the other girls—both Catholic and Protestant. Her recollections and those of fellow student Lucy Thaxter paint a picture of a strict but not overly harsh regime of study, prayer, and modest meals.

This was not, however, the impression of many Charlestown residents. Stories of dark deeds, cruelty, and ghastly rituals behind the convent walls circulated in Charlestown streets and taverns, chiefly instigated by a former student named Rebecca Reed. Reed was not Catholic, but as an orphan with few resources, she expressed interest in life as a nun and was accepted to Mount Benedict as a charity scholar in 1831. After six months, she chafed at the school's discipline and "escaped" by jumping over a wall, although school officials later said they would have

The Ursuline Convent

gladly seen her out through the front door. Reed, an obviously troubled woman who craved attention, began spreading lurid stories about the convent. She told of how hard penance cut short the lives of tender young novices, that nuns were ill fed and ill treated while the bishop and Mother Superior lived like royalty, and that, overhearing plans to kidnap her and ship her to a Canadian convent, she resolved to escape soon "or it would be forever too late." Those threatened by the growing number of Boston Catholics spread her tales of nunnery travesties.

Just a few days before Louisa came to Mount Benedict, Sister Mary John (Elizabeth Harrison), the school music teacher, who had been at the school twelve years, apparently suffered a nervous breakdown, brought on by overwork. After a bout of hysterics, she fled the convent and showed up at the neighboring home of Edward Cutter in a desperate state. After intervention by the bishop and her brother (and perhaps after she finally got some rest), the calmer and now contrite sister was convinced to return to the convent, where she seemed to recover from her "hysterics." News of her "escape" spread, however, and seemed to confirm Reed's stories of women kept against their will. A short article appeared in the *Boston Mercantile Journal* on August 8, with the headline "Mysterious," hinting that a young lady who had "made her escape" from the convent was being held against her will. Placards began to appear in Charlestown calling on the town's selectmen to investigate the "mysterious affair"; if not, "the Truckmen of Boston will demolish the nunnery."

If Reed's stories and Sister Mary John's "escape" were the tinder for the events of August 11, 1834, the spark may have been provided by three sermons given days earlier by Lyman Beecher. "The principles of this corrupt church are adverse to our free institutions," declared Beecher, president of the Lane Theological Seminary. While Catholic schools taught Protestant students, "the children of the subjects of the Pope were left to roam in ignorance, many of them incapable of either reading or writing." Beecher was, moreover, indirectly responsible for Louisa Goddard's misery. Her father violently opposed Beecher's orthodoxy, so in a rousing spirit of antagonism he decided to send his daughter to the very place Beecher warned against. Louisa didn't care about politics, she was more concerned with making new friends at school. She was vaguely aware of strange stories about the "nunnery" and that a mob might be out to destroy it. Her father, like the fathers of other daughters, paid little heed to the threats of violence.

More than anti-Catholic sentiments were at work here. The "truckmen" (an early version of the Teamsters, whose loosely organized crews hauled two-wheel carts), firefighters, brickmakers, sailors, and other laborers despised Boston's wealthy as much as they did the Irish. The convent and its wealthy students triggered resentment of the upper classes as well as hatred of the poor Irish. Salem State College professor Nancy Lusignan Schultz, author of *Fire and Roses,* a landmark book on the convent fire, speculates that the independent and self-sustaining Ursulines, particularly their "saucy" Mother Superior, posed a challenge to male authority: "Ironically, had Moffatt not been as ambitious, capable, and visionary in building Mount Benedict, the school and the Ursuline order might have survived nineteenth-century Boston. Her personal success was in some way the institution's downfall."

Certainly the Mother Superior made no effort to diffuse tensions. She was rude to neighbor Edward Cutter when he came to look in on Sister Mary John. Only after much persuasion did she allow Charlestown selectmen on the morning of August 11 into the convent to investigate the wild stories circulating about strange, unholy rites and dead bodies. After three hours of inspection, the selectmen, satisfied that all was in order, promised to circulate their findings. By the time they issued a statement, it was too late. Likewise, Cutter delivered a statement to a local paper and it was printed on August 12, when the convent's ashes were cooling.

Just after sundown the night of August 11, men began to gather at the gates of the convent and were watched with increasing apprehension by the nuns and the students. By 9 P.M. the crowd had swelled to several hundred; many of the men had marched over the bridge from Boston, shouting as they crossed the river. "What shall I call it?" Louisa later recalled. "A shout, a cry, a howl, a yell? It was the sound of a mob, a voice of the night, indeed that made it hideous." Lucy Thaxter remembers seeing "a dense black mass, apparently moving up the avenue toward the convent and the sound of their prolonged huzzas came upon my ears like yells of fiends. Never shall I forget that sound."

The group was led by the 29-year-old John R. Buzzell, a New Hampshire native and father of five children, who came to Charlestown to work as a brickmaker. Described as a broad-shouldered, imposing man, he often bragged of his wrestling skills, perhaps a cover for frustration at his inability to make a better living after an economic downturn in 1833. On the night of August 11, according to later court testimony, he stopped off at a store and, over a glass of gin and molasses, declared to a companion that he was ready to be the first man to break into the convent. At his instigation, men began to gather at the bottom of the hill below the school. They began lighting bonfires outside the gate, perhaps to signal other rioters or to call in volunteer firemen. About 9:30 P.M., fifty to a hundred men marched up the road to the door of the convent. Upstairs, a rattled Sister Mary St. Augustine began waking students, saying, "Girls, girls, there really is no danger—but you had better dress yourself." The Mother Superior, although agitated, remained stern and very much in charge. "With the courage of a man," as Louisa put it, she marched to an upstairs window and stared down in the darkness to the mob below.

"What do you want?" she demanded as the crowd fell silent to hear her. "We want to see the nun who ran away," the men shouted back. Haughtily, St. George said that the selectmen had thoroughly investigated the situation; then, displaying more courage than tact, she ordered the men to leave. "Disperse immediately, for if you don't, the Bishop has twenty thousand Irishmen at his command in Boston, and they will whip you all into the sea!"

Her challenge was like red meat to wolves. Although the men seemed to withdraw down the hill, within an hour they came back with greater force than ever. Buzzell later claimed the men held a meeting and that a "motion" was unanimously passed to tear down the convent. Now, waving torches, their faces painted in some attempt at disguise, they marched toward the convent, shouting, "Down with the Pope! Down with the Bishops! Down with the Convent!" They were determined, as Buzzell put it, to "clean the establishment out." At some point, Lucy Thaxter recalled, one of the laywomen begged the men not to hurt them, saying, "We are only women and children and there is no one to protect us." "That's all we need to know," a man replied with glee.

From the convent dormitories, the students and nuns heard the crash of breaking glass and splintering wood as men began to storm the convent. In the ensuing chaos the sisters gathered up the girls—many still in their nightclothes—and herded them into the convent's garden. A rattled Louisa could hardly believe what was happening. Earlier, noticing the glow from the bonfires and seeing the arrival of the fire engine from Charlestown, she assured a fellow student, "They will certainly help us, those firemen. They certainly will drive the mob off." But she and the student watched helplessly as the fire engine disappeared.

The seemingly fearless Sister St. George now led the exodus. She knew she would be killed if the rioters found her. Yet, to Louisa's admiration and amazement, "her eye never quailed and neither hand nor voice trembled." Other nuns also acted with dispatch and courage. Two tried to save the convent's precious tabernacle; they carried it out of the house into the garden and hid it in a patch of asparagus gone to seed. It was later discovered and savagely kicked down the hill. And when one nun fainted, she was carried by two others down two flights of stairs into the garden. The girls, meanwhile, scampered through the garden to the fence at the far end, where they cowered and tried to stay quiet. Looking back, Louisa could see by lantern light rioters ransacking the building and tossing furniture, books, china, and musical instruments out of the windows to great cheers and hurrahs. Then faint flickers of fire appeared and soon the windows of the convent began to glow. "These were the most horrible moments of all that horrible night, and the noise was aggravated by the increasing roaring of the fire, which together with the brilliancy of the light and the pungent smell of smoke, threw the poor women and children about me into a stronger agony of terror than ever; the harder to bear because it had to be suppressed."

Inside the convent the rioters, having ransacked the building, were determined to destroy what was left. Wearing the bishop's robes, which he had donned "in a spirit of deviltry," Buzzell urged on the destruction. Crosses, Bibles, religious vestments, and altar ornaments were hurled into the roaring flames. Sixteen-year-old Marvin Macy acted the part of a mock auctioneer, shouting, "Sold," as he tossed books into the fire. The rioters also burned the barn and the bishop's residence and even broke into the convent's cemetery and disturbed the bones of the dead. "They desecrated the mortuary chapel and the tombs where our dead Sisters were sleeping their last sleep, they scattered the ashes of the nuns to the winds and they carried their blind fury as far as to remove the teeth of some of the dead bodies," Sister Mary Joseph lamented.

Hiding in the garden, the nuns and girls now feared for their lives. Lucy Thaxter and another girl, grasping hands, gazed "into each other's face in silent horror. Could such things be? Were they men with hearts of fathers and husbands beating in their bosom?" When they heard a noise from the other side of the fence, the girls were nearly beside themselves with terror. But the voice belonged to Mr. Cutter, who quickly and quietly helped the women and children over the fence and took them to the relative safety of the home of the Joseph Adams family about a half-mile away. There, exhausted and overwrought, the children and sisters collapsed. A young novice, stricken with tuberculosis and who would die in a few weeks, was carefully laid to rest on a sofa.

The Ursuline Convent in flames

As the women attempted to collect themselves, Adams came to the door and said gently, "Ladies, if any among you wish to take a last look at your convent, follow me." The sisters and Mother Superior slowly climbed the stairs to the top floor of the house where they could see their school, now a mass of flames. After ten minutes they could stand the sight no longer and returned to the parlor, where they knelt in prayer. "My God, forgive us our sins as we forgive those unfortunate fanatics," Sister Mary Joseph wept.

The blaze was visible for miles around, and yet no authorities appeared to quell the blaze or halt the violence. When fire companies from Boston and Charlestown arrived, firefighters made no effort to douse the flames. "Enough engines responded to have flooded the whole hill, but only one, the North End, No. 2 of Boston, went to the nunnery, the others stopping at the foot of the hill," Buzzell recalled years later. "Not one of them played a stream during the entire conflagration, while many of the firemen were aiders and abettors of the mob. . . ." The North End company apparently found the situation beyond control, and its commander ordered the engine and men to retreat down the hill. At one point, a Charlestown selectman did attempt to reason with the mob, but he soon gave up and went home to bed. More astonishing, while only about 50 to 200 men participated in the actual ransacking, thousands watched the flames without intervening. Buzzell put the crowd at 4,000; other accounts put it at 1,000 to 2,000.

In the morning the nuns and children parted forever. Louisa made her way home, to the great amazement of her father, who had not yet heard of the convent burning. Meanwhile, rioters returned the next evening to burn the orchards and fields surrounding the convent, even as major figures in Boston gathered at Fanueil Hall to condemn the attack. Mobs, roaming Boston's streets, also burned a Charlestown shanty that housed Irish families, and another mob attempted to burn the Catholic Cathedral in Boston.

Eventually, thirteen men were arrested for the convent fire; a trial was held in December 1834. Buzzell was not worried. "The

testimony against me was point blank and sufficient to have convicted twenty men, but somehow I [provided] an alibi, and the
jury brought in a victory [i.e., verdict] of not guilty, after having
been out for twenty-one hours," he said during his trial. Another
rioter, Benjamin Wilbur, a fireman attached to a company that
had been to the fire, confessed on his deathbed that actions were
planned two weeks earlier. He even implied that plans were actually hatched in the engine house, according to a history of the
Boston archdiocese. The only person convicted was the teenager
Macy, who was sentenced to life imprisonment. His harsh sentence outraged many, including the Mother Superior and Bishop
Fenwick. They were among those who signed a petition to have
the sentence commuted, which it was. The only sentence served
was self-inflicted. One of the rioters, overcome with guilt, committed suicide by slicing his own throat; communion wafers
were found in his pocket.

The Catholic archdiocese sued the state for damages—estimated at $50,000—due to the lack of intervention by authorities.
The suit went on for decades, but the church never recovered a
cent. While Boston leaders roundly condemned both the fire and
the firefighters who stood by, violence against Boston's Irish
Catholics continued. In 1835 Rebecca Reed's sensational account,
Six Months in a Convent, was published and was an instant success, although Schultz suspects it was written largely by an anti-
Catholic committee. The introduction shrugged off the convent
fire by asserting: "If Protestant parents will resolve to educate
their daughters at Protestant schools and patronize no more Nunneries, then no more Nunneries will be established in this country, and there will be none for reckless mobs to destroy." Mother
Superior St. George wrote a response, and others attacked the
veracity of Reed's account, but tension continued.

The tension culminated on June 11, 1837, in an event that permanently sullied the reputation of Boston's volunteer fire department. Newspaper accounts of the circumstances that led to "the
Broad Street Riot" differ widely, and even accounts written
decades later reflect the passions or background of the author.

Boston fire historian Arthur Brayley tilts the blame for the violence toward the Irish participants; a Boston church history account lays the burden of guilt squarely on firefighters.

This much is known: On a hot Sunday afternoon, Engine No. 20 had returned from a fire, when firefighters encountered an Irish funeral procession of several hundred people. A fireman, George Fay, apparently jostled or was jostled by some of the mourners and insults and fists began to fly. Fay testified later that he was "going along East Street, an Irishman shoved me from the side-walk, and I fell to the ground. I asked him what he meant. He said that is the place for you and struck me." A church history account asserts that the 19-year-old Fay "who lingered longer than his comrades over his cups, came swaggering through the crowd, smoking a cigar." However it started, Fay and a group of Irish men decided to settle things with their fists.

Hearing the ruckus, firefighters from Engine 20 rushed over, and the fight became a brawl. The Irish, who outnumbered the firefighters, quickly rousted the Yankees and the funeral procession went on. But the routed firefighters now spread a call to arms with the cry: "The Irish are revolting." Some rang the fire bells, perhaps to get help, perhaps to draw attention to the violence. One man ran to another firehouse, shouting, "The Irish have risen upon us, and are going to kill us." A second engine, No. 9, rushed to the area, hitting or very nearly hitting people in the funeral procession. An investigation later never concluded whether this company knew it was coming to a fire or a fight.

The brawl now became a riot. Firefighters chased the Irish funeral goers to Broad Street, which was then a residential area where many of the city's Irish lived. Pent-up hostilities were released in a frenzy of fists and insults. Rumors quickly spread of deaths on each side. The mob now turned its anger on the buildings; men broke into the tenements, grabbing whatever they could find, as occupants rained hearthstones down on them. Eventually, the now overwhelmed Irish, including women and children, fled the street, leaving every possession behind. Rioters smashed windows and hurled household goods into the streets;

some Irishmen were dragged from cellars and beaten. "A gang of stout boys and loafers who had followed the fireman at such distance they might be protected from the dangers and at the same time participate in the mischief of the affray, attacked the houses of the Irish in the rear of the scene of the combat, tearing to pieces and destroying everything wantonly and recklessly," according to Brayley. A former fire captain was severely beaten and was about to be tossed off a wharf when one of his Irish attackers stopped his comrades from this final act of bloodshed. Using bats, axes, and hatchets, the brawlers fought with the blind fury of those who have forgotten what they are fighting for but believe they must fight on.

An astonishing 10,000 to 15,000 people, including 700 firefighters, were said to have been involved in the riot. The violence continued until Mayor Samuel Eliot called out various companies of state militia and order was restored, leaving Broad Street ankle-deep in feathers and straw from the ransacked houses. About 29 families, including 122 people, lost their homes and their possessions. No deaths were reported, although scores were wounded, and some people may have died of their injuries later at home. Some Boston newspapers expressed outrage at the rioting firefighters, suggesting that they were "actuated by the vindictive and destroying spirits of fiends." Other papers rushed to their defense, noting, "We cannot but allow much for the exasperated feelings of men who understood their friends had been murdered." Thirty-four "bleeding Irishmen" were arrested. A grand jury eventually indicted fourteen of them and four Yankees for rioting. Unsurprisingly, the Irish were convicted and the non-Irish were acquitted. "The plebeian firemen emerged unscathed and victorious," historian Jack Tager concluded.

The romantic image of the volunteer fireman was, however, finally shattered. Men regarded as selfless saviors were now viewed as rioters and bullies. Members of Engine 7, which had been involved in the riot, were unrepentant and published an angry defense in the Boston papers, infuriating Mayor Eliot. He dismissed the company and announced a reorganization of the fire department. An ordinance passed on August 10, 1837, called for the hiring of full-time, salaried firefighters; banned "races returning from fires"; and gave the mayor and the city council authority to make fire department hires. While citizens continued to be designated as "call" firefighters, the era of the independent, volunteer fireman was coming to an end.

The shell of the Ursuline school remained on Mount Benedict for more than forty years. The Catholic Church refused to sell it, seeing it as a monument to religious hatred; residents picnicked among the ruins. After several attempts, the Ursulines decided not to establish a new school; most of the nuns and St. George left, never to return. Buzzell later became a New Hampshire legislator. He lived to nearly age 89 and died a respected farmer. He showed little remorse for his actions and went to his grave insisting that tales of abuse within the convent walls were true. Rebecca Reed died at age 26 of tuberculosis.

In the decades that followed, Irish immigrants and their children came to dominate Boston's politics and cultural life. The section of Charlestown where the convent stood was annexed to

View of Somerville from the ruins of the convent

the new town of Somerville in 1842. Eventually, the remains of the nunnery were carted away, the hill leveled, and the dirt used as landfill for narrowing the nearby Middlesex Canal. The area, today bordered by Broadway and Mystic Avenue, was soon covered with houses, although Somerville old-timers continued to call it the "nunnery grounds." Bricks from the ruins were used to form the arch of the front vestibule of the Cathedral of the Holy Cross in Boston's South End.

Despite her father's best efforts, Louisa Goddard became an assertive woman and a writer. She married renowned geologist Josiah D. Whitney and traveled with him across the country. As California's state geologist, Josiah Whitney was the leader of one of the first groups of white men to scale the highest peak in the continental United States; Mount Whitney was named after him. Eventually, the couple returned to Massachusetts, where Mr. Whitney taught at Harvard.

Almost forty-two years to the day after she was told she would be attending the Mount Benedict school, Louise Whitney and her husband passed by the ruined site in a carriage and saw workmen finally clearing away the debris. Her husband—a man quite unlike her father—urged Louisa to write down her memories of that ghastly night. She concluded her 1877 book, *The Burning of the Convent*, with an example of the kind of behavior her father had hoped to curtail. After coming out in society, she was attending a ball when a "middle-age dandy" with a pretentious eyepiece sought to impress her with tales of his youthful spirit.

"My dear, I even formed a party of gentlemen who followed those rioters into Charlestown; we were curious as to what they would do, and seating ourselves on the grass at a respectable distance, we watched until they completed their destruction," he told her with relish.

"And did you do nothing to prevent it?" Miss Goddard asked indignantly.

"Why—a—no. There didn't seem to be anything we could do. Ah—there is our waltz at last. Allow me." But Louisa coolly drew back from the man's arm.

"Excuse me but I don't like a feeble partner. I like to be held up firmly when I waltz and that arm must be weak indeed which had not strength to uplift itself in defense of helpless women and girls," said the young woman who would never forget how prejudice could burn out of control.

BUILT IN BOSTON

The Soul of the Old Masheen

The once common expression "dressed up like a fire engine" seems odd today—with good reason. Modern fire engines and ladder trucks, while sleek and shiny, are inevitably red and designed for speed and efficiency rather than for show. To understand the expression, imagine yourself in early 1840s Boston. Clanging church bells cry out "Fire." Down the street, running at top speed, come firemen pulling a fire "engine," a machine that used manpower as its only source of energy. The men may be sweaty and dirty, but their machine gleams with components of brass and polished wood, embellished with gold or silver leaf. As they race by, admiring passersby get a glimpse of the intricate scrollwork on the body, the patterns painted on the wheels, and the swinging lanterns and bells. As the men reach the fire, followed by excited boys who have chased them, they quickly attach a hose to a water source—a water plug or hydrant. As a couple of nozzlemen aim the hose toward the flames, others line up along the long handles of their engines, which also served as the brakes. With fierce strength, they begin to pump. Water is sucked into the machine, and with pressure created by the pumping motion, a stream of water is sent soaring toward the roof of the burning building. Every ten to fifteen minutes, new men step in to take over the arduous pumping, giving each of the forty to a hundred men in the company a stint at the machine.

Another fire company clatters into sight. These men, however, look a bit chagrined; they see that another company has beaten them to the scene. Still, they can think smugly, they might be late but *their* engine, with its elegant design, painted murals, and gold leaf, is a far prettier piece of work than that of their "competitors." Not only that, but they are sure that, given the chance, they could send their stream higher than that other company—maybe higher than any other company in New England.

This scene was repeated throughout Boston whenever the dreaded cry of "Fire" rang out. The days of the early nineteenth century were glory years for volunteer "fire laddies" and the romance of "running wid der masheen," as a popular expression put it. Volunteer firefighters were celebrated as embodying the very spirit of America in poetry, art, even theatrical productions that lauded the brave, fearless citizens who put their lives selflessly at risk. The popular print partnership of Currier and Ives frequently featured firefighters and firefighting scenes. Nathaniel Currier, also a firefighter, posed himself in the print "Always Ready," in a series

"Always Ready," an 1858 print from the Currier & Ives "American Fireman" series. Reportedly it shows Nathaniel Currier, who was a volunteer fireman.

of lithographs on "The American Fireman." In an 1809 address to the Charitable Fire Society in Boston, Alexander Townsend echoed the general sentiment when he declared, "Volunteers in the service of beneficiaries are the glory of civilized life."

The centerpiece of any fire company was its "masheen." As an unnamed Boston newspaper, quoted in the *Fireman's Journal* of 1855, put it, "With a fireman's love of 'the machine' he grasped the rope and managed his way through a raging sheet of flame, burning his hat, hair and clothes." This love grew into a passion of astonishing depths. Firefighters lavished attention on their engines with an ardor that fire historian Amy S. Greenberg contends bordered on mania—even fetishism. For many, the engine was their "lady," lavished with gifts of ornamentation and decoration and emblazoned with mottoes like, "We Come, We Conquer," "Fear Not, We Come," and "We Will Try." Men dressed their lady in bright, gaudy colors, adding scrollwork, maybe a mural or two, gold leaf, and silver lanterns. (Red was not generally used until a vibrant red paint was developed in the late 1840s. It was expensive, therefore fire companies then *had* to use it on their engines and eventually "fire engine red" became a common saying.) Firefighters swapped tales of comrades who deserted brides on their wedding night at the sound of an alarm. Even if these stories were exaggerations, an insult to a company's machine was taken as seriously as an insult to a wife and could send fists flying. Moreover, the public loved fire engines, regarding their companies' spirit and engines as manifestations of community pride.

No manufacturer in New England was more famed for the power and beauty of its hand-pumped engines than the William C. Hunneman Company of Boston. Perhaps it was not surprising; the founder, William Hunneman, was an apprentice to one of the great masters of metal craft in colonial America—Paul Revere. During the glory days of the hand pumpers, "the Hunneman was king, acknowledged as the finest engine money could buy," declared fire historian and Hunneman expert Edward R. Tufts. From 1792 to 1883 the company manufactured 745 fire engines, many of which remained in service for more than fifty

Hunneman's notice for his patent

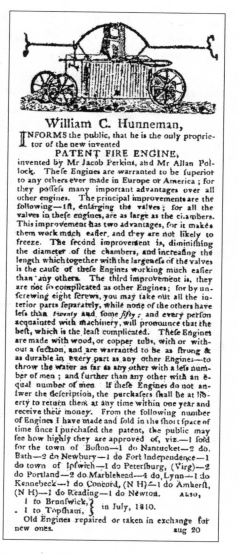

William C. Hunneman,

INFORMS the public, that he is the only proprietor of the new invented

PATENT FIRE ENGINE,

invented by Mr Jacob Perkins, and Mr Allan Pollock. These Engines are warranted to be superior to any others ever made in Europe or America ; for they possess many important advantages over all other engines. The principal improvements are the following—1st, enlarging the valves ; for all the valves in these engines, are as large as the chambers. This improvement has two advantages, for it makes them work much easier, and they are not likely to freeze. The second improvement is, diminishing the diameter of the chambers, and increasing the length which together with the largeness of the valves is the cause of these Engines working much easier than any others. The third improvement is, they are not so complicated as other Engines ; for by unscrewing eight screws, you may take out all the interior parts separately, while none of the others have less than *twenty* and some *fifty* ; and every person acquainted with machinery, will pronounce that the best, which is the least complicated. These Engines are made with wood, or copper tubs, with or without a suction, and are warranted to be as strong & as durable in every part as any other Engines—to throw the water as far as any other with a less number of men ; and further than any other with an equal number of men. If these Engines do not answer the description, the purchasers shall be at liberty to return them at any time within one year and receive their money. From the following number of Engines I have made and sold in the short space of time since I purchased the patent, the public may see how highly they are approved of, viz.—1 sold for the town of Boston—1 do Nantucket—2 do. Bath—2 do Newbury—1 do Fort Independence—1 do town of Ipswich—1 do Petersburg, (Virg)—2 do Portland—2 do Marblehead—4 do Lynn—1 do Kennebeck—1 do Concord, (N H)—1 do Amherst, (N H)—1 do Reading—1 do Newton.   ALSO,

1 to Brunswick, } in July, 1810.
1 to Topsham, }

Old Engines repaired or taken in exchange for new ones.   aug 20

years. Their handsome designs and well-crafted components were shipped all over the country, and engines were often sold from one fire department to another. While New England had other makers of hand-pumped engines, such as the Thayer Company of Boston, Howard-Davis of Boston, the Leslie Company of Newburyport, and the Jeffers Company of Rhode Island, and other local craftsmen who would turn out one or two engines a year, Hunneman built more hand pumpers than any of them. Yet some makers of hand pumpers continued to build machines into the twentieth century, while Hunneman's celebrated expertise and attention to detail proved its undoing. Tradition-bound Boston firefighters resisted nearly every new innovation. But once change was forced upon them, they gradually accepted it and made it a tradition—until the next innovation came along. A pattern of innovation-resistance-acceptance was repeated through the nineteenth century, and the Hunneman Company failed to keep up with that cycle.

Whereas Boston had its share of foot-dragging firefighters, the city never experienced the violence of other cities, such as Cincinnati, where volunteer firefighters rioted in response to the introduction of steam engines. Rather, Boston had a long history of nurturing technological developments that eventually won over those committed to battling blazes. During colonial days the town was the first in the colonies to obtain new "injines," primitive contraptions designed to shoot a stream of water into flames. After the 1653 fire, Boston ordered its first fire "injine" from a Mr. Jenks (also spelled Jynks), an innovative ironmaker who had already come up with modifications to sawmills and scythes. This apparatus probably was a kind of syringe pump that, supplied by bucket brigades, would "squirt" water into fires. Similar machines were used in London. Some historians wonder if the Jenks machine was actually delivered; there's no evidence that it was effective (or even used) in the fire of 1676, for example.

In 1678 Boston ordered from London a more sophisticated firefighting device: the first hand tub, soon dubbed "ye Engine by ye Prison," because it was stationed by the municipal prison. By modern standards the "engine" was primitive; it consisted of a wooden tub, a direct force pump, and a small hose. The contraption would be carried to a fire, and the bucket brigade would fill the tub while others pumped the handles to create pressure for the hoses. The device proved so useful that more were soon ordered; by 1720 the town had six of them and by 1760 it had

nine, some made by the innovative engine designer Richard New-
sham of London. Possession of the finest available fire defense
became an early Boston tradition.

For a time, in Boston's pursuit of state-of-the-art firefighting
devices, William C. Hunneman would lead the charge. Born on
July 10, 1769, to Nicholas and Elizabeth Cooper Hunneman,
William Hunneman became an apprentice of Paul Revere, learning from him the craft of copper, brass, and silver design and production. Revere might be best known for his patriotic midnight
ride—owing to the poetic license of Henry Wadsworth Longfel-
low—but during his life he was as renowned for his skills as a sil-
versmith. After soaking up knowledge from the master,
Hunneman worked at a brass foundry, operated by Martin Gay,
which made utensils, wind vanes, and other products. Gay, however, was a Tory and fled the Americas for England after the Revolutionary War, leaving Hunneman alone in the business. By the
century's end, the enterprising Hunneman saw a future in a new
kind of product.

Purchasing a patent from inventor Jacob Perkins in 1802, Hunneman decided to go into the fire engine business. He modified
the Perkins design to create a more sophisticated pumping system and was soon selling fire engines to fire companies in Massachusetts, Rhode Island, and Maine. In a notice published in
1810, Hunneman boasted:

> These engines are . . . warranted to be as strong & as durable
> in every part as any other Engines—to throw the water as far
> as any other with a less number of men; and further with an
> equal number of men. If these Engines do not answer the
> description, the purchasers shall be at liberty to return them at
> any time within one year and receive their money.

In 1808 Hunneman moved his rapidly expanding business from
the North End to Roxbury. For the next forty years he ran the
company—eventually with his son William Jr.—and produced
more than 300 machines. Every piece of metal was hand-forged

and hand-finished to specification. Not only were machines carefully numbered, beginning with No. 1, but they were also given names, meticulously recorded by the company. Out of Hunneman's shop rolled the Davy Crockett, Always Ready, Cataract, Victory, Deluge, Neptune, Protector, Fire-King, Tiger, Conqueror, and Niagara, to name but a few. Many went to New England fire departments, but Hunneman shipped machines all over the United States and the rest of the world, including Chile, China, Turkey, Cuba, and the Philippines.

Hunneman continued to update his designs. In 1823 the company produced its ninetieth engine, which became "Torrent," No. 16, for Boston. It was the first engine equipped with a suction device that allowed the engine to draft water directly from a water source. The innovation was made possible by the development of sturdy, reinforced hoses that withstood the vacuum created by the pressure of suction. The bucket brigade was relegated to history, and a new piece of equipment, a wagon for hauling hoses, was created; Hunneman was ready to produce those as well. Hunneman machines also began to feature a lightweight, crane-neck design, making them easier to maneuver in city streets. The machines were further embellished with brass bells or a brass eagle on top; once delivered, fire companies would add additional scrollwork or cover the sides with paintings, often of a woman and sometimes of a nude woman. A newspaper article from the 1850s, found decades later among the Hunneman records, described an extensive operation that handled every part of fire engine production, from the metal and iron work to the final paint job. "A visit to these works is a rich treat to all who run 'wid der machine,' or take any interest in firefighting apparatus," the article concluded.

Prices reflected the increasingly sophisticated designs. In 1807 a hand tub sold for $420, by 1816 the price was $355 to $655; and in the 1840s prices ranged from $950 to more than $1,000. The company even added precious metals to its machines; the components of the Lafayette were silver-plated before it was delivered to Boston on January 6, 1837. Firefighters promptly nicknamed it

"Silver Hinges." The most expensive Hunneman was delivered by mule train to San Francisco in 1856; the "Mayor Brannon" was first sent to New Haven to be heavily silver-plated, and later ornamental pictures were painted on its sides and stern box. Its four buckets each featured a rendition of the four seasons. The price tag was $12,000—a fortune in those days. But no price was too high in the constant push to outdo the other companies.

Pride goaded fire companies to obtain the prettiest engines they could afford. But competition involved more than appearance. An even greater incentive to competition—one that dated back to the colonial era—was response time. Back in 1740, Boston councilmen—to encourage "the respective companies belonging to the several Fire Engines in this town and to stimulate them to their duty in extinguishing of fire"—offered a £5 bonus to the company whose engine "shall first be brought to work upon any house or building that shall be on fire." This practice of rewarding those companies arriving first to a fire, a custom that soon spread through the country (and was often encouraged by fire insurance companies), brought out the best—and the worst—in men. Soon the bonus became a point of pride as well as financial gain. "Each

"St. Louis," Hunneman Engine No. 574, shown in front of the Merchants Exchange Building in Boston. This model shows the kind of artistic embellishment fire companies often added to their engines.

man strove mightily to make *his* company the best one in town. That meant humbling the other companies: in racing to a conflagration, in getting a stream on the blaze, in amount of water pumped, in appearance of fire apparatus. At first this was a friendly rivalry, such as stimulates endeavors in any field. But in time the competition became fierce," writes Robert S. Holzman in *The Romance of Fire Fighting*. Eventually, "the men spent more energy in battling with other companies, in many instances, than they did in fighting fires." The expression "plug ugly" stems from the ferocious battles to be the first to attach a hose to a fire plug or hydrant. The first company "plugged" was deemed the "first" at the fire, and various subterfuges were used—including attempts to hide the hydrants—to get the coveted "first." Sometimes buildings blazed away as firemen fought for the honor of being the ones to put the fire out.

An 1860 photo of the "Torrent Six" (built in 1829) at the Roxbury Fire Department. Samuel H. Hunneman is on the left of the brake arm, and John C. Hunneman is on the right. The other men not in uniform are employees of the Hunneman Company.

Many times engines had to be connected in a series of relays to bring water from a far-off water source; water was pumped from one machine to the next. It was a serious blow to a company's self-esteem to be "washed." When one company could pump more water into the next engine than that company could pump out again, the excess water overflowed from the "washed engine." Being washed meant your company wasn't as strong as the others in the relay.

City officials also took advantage of the competitive spirit among the fire engine manufacturers. In 1831 Boston wanted a new engine for a newly organized company. Officials proposed that the Thayer Company and Hunneman each manufacture an engine of the same capacity and pit them against each other before the city board of fire engineers. The companies agreed. The Hunneman engine and the Thayer engine squared off near the corner of Dover and Washington Streets. In a variety of tests the Hunneman "proved to be superior in every way." So Hunneman No. 140 was returned to the company, painted, ornamented, and christened the "Eagle." In all, Boston purchased forty-six Hunneman hand pumpers.

Fire companies also turned rivalries into more healthy athletic competition. Beginning about 1849, companies staged musters to see which company could pump water faster or shoot streams farther. Such musters—coming before the days of modern baseball and football leagues—were the chief sporting events of the era. Huge numbers of spectators would gather on Boston Common to root for their favorite "team." But competition continued to have ugly consequences as many men turned fire houses into social clubs for drinking, playing cards, and roughing up other companies. Firefighters were sometimes seen as rowdy street toughs, as ready to pick a fight as to fight a fire—a characterization borne out by the 1837 Broad Street riots. David D. Dana, a firefighter and the author of a fire history book published in 1858, felt obliged to challenge the public's perceptions: "The great body—the mass of firemen—suffer from the wrong-doing and the vices of a few of their members. From a careful investigation, it will be found that the firemen as a

*A firefighting
contest between
"smoke-eaters" on
Boston Common,
1851*

body will average, in point of respectability and worth as producers and artisans, in point of moral character as citizens, and indeed in all relations of life, equal to any organized class of men."

Camaraderie and tradition could, however, close ranks against others invading this Yankee male bastion of civic pride. Dana, for example, also railed against the "introduction of foreigners" to departments. "We are good neighbors, hospitable, kind, social, generous; but must deny their right to interfere with the management of the affairs of our American homestead. When foreigners become members of the fire department, from that moment, as has been the constant experience, dissensions [sic] arise." The foreigners Dana had in mind were the Irish, and the barrier he envisioned would not hold for long. The Irish started filtering into Boston's fire department after the Civil War; they joined in even greater numbers after the 1872 fire. In 1906 the Irish-American mayor John E. "Honey Fitz" Fitzgerald (grandfather of President John Fitzgerald Kennedy) appointed the first Irish-American fire chief, John A. Mullen.

Love of tradition could also turn into a stumbling block when innovations emerged. Companies often refused to give up old ways, even for more effective firefighting. For example, fire historian Paul Ditzel records that in 1819 a Boston fire company decided not to waste time in hooking up to a water source. They filled their rig with water so they could begin pumping as soon as they got to a fire. But Boston officials called foul, saying the stored water gave the company an unfair advantage over others. An idea standard in today's fire engines was banned in the name of competition. Early firefighters even opposed the introduction of hoses to draw water from a source to a hand tub; they preferred the bucket brigade routine. As a New York fire official later declared, "I esteem the fire department as one of the most difficult departments into which to introduce any changes or innovations, even though they may be exceedingly important, and commend themselves to the common-sense judgment of every man."

Another innovation vociferously opposed was the introduction of the steam fire engine, first built in England in 1829. This

No. 678, "Somerville No. 1," the first steamer the Hunneman Company built

machine burned wood or coal to build up water pressure and force out a stream of water. Historians have not failed to notice the irony—fire would be used to fight fire. Two Ohio men built the first successful steam engine in this country in 1852 (after an early version broke down miserably in New York). Less than three years later, Boston bought its first steam engine, the Miles Greenwood. It was, unfortunately, large and ungainly and therefore difficult to maneuver through Boston's narrow streets. A city official who had pushed for purchase of the new steam fire engine later complained that although "the engine was superior in efficiency, I believe, to any steam fire engine since used, so strong was the opposition of the fire department to it, that when exhibited to test its power (firefighters) cut the hose and injured it in such a way as to prevent its operating satisfactorily, and it was finally left to rust and ruin, and sold for old material; and this was not at all from any defect in the engine, but because the department was determined not to have it introduced."

Firefighters regarded the new engines as an affront to their ability. The idea that a choking, sputtering machine could compete with manly men was a blow to the smoke-eaters' pride, and many refused to have anything to do with the newfangled contraptions. Moreover, steam engines were heavier than pumpers, requiring that they be drawn by horses, not men. Many firefighters were appalled at the idea of turning part of their station into a smelly stable. Ohio steam engine designer Alexander Latta complained in a letter that all he wanted was the city of Boston to "give it a fair chance." Latta even told a friend despairingly, "I sometimes fear that I shall never live to see this grand idea brought into the service of the world." Naturally, steamers and hand pumpers went head to head in contests and, as the Boston *Fireman's Friend* reported, "the Steamer came off, as usual—second best." In Cincinnati in 1853, the Union fire company took the city's most powerful Hunneman, Ocean Company No. 9, and challenged the new machine in a hand-to-steam contest. But the best the volunteers could pump was 200 feet, while the steamer easily reached 225 feet. Men even competed with horses in "running with the machine" and some-

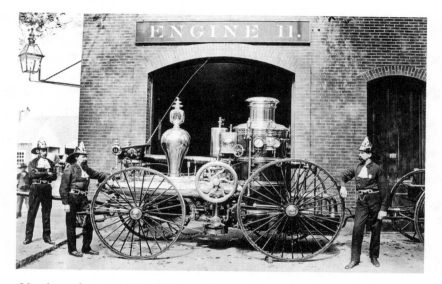

Members of Boston's John S. Damrell Engine Company No. 11, in 1866

times humans outran equines. Still, in December 1858, Boston pur-
chased two steamer engines: the Eclipse No. 6 (built by Silsby
Mynderse and Company of Seneca Falls, New York) and the
Lawrence 7 (built by Bean and Scott of Lawrence, Massachusetts).
By 1860 all the city's nine engine companies had steam
engines—many of them built by the Amoskeag Manufacturing
Company of Manchester, New Hampshire—and the nine hose
companies and three ladder companies were all horse drawn.
Eventually, the new machines were polished with the same loving
care as the old pumpers, and many were embellished with hand-
lettering and scrollwork. Boston sold its last hand pumper, the
Hunneman-made S. R. Spinney, to San Francisco in 1861. While
still looking back nostalgically on the days of "running with the
machine," the public was now also enthralled by the speed and
drama of a horse-drawn fire engine. "The sight of a steam fire
engine drawn by three galloping horses the engine belching smoke
and whistling furiously, was justified cause for thrilling excitement," noted Boston fire history expert William Werner.

The transition to steam power was, in retrospect, inevitable.

The new engines shot streams farther, meaning men could stand at a safer distance from fires. Fewer men were required to operate them. And while hand-pumped engines could often get water on a fire faster once they had arrived at a scene (the steam engine, after all, had to have its fire lit and stoked), men were no match for machine in stamina. Moreover, steam engine operators developed ways to get their fires going as soon as the engines were called out; many were pulled rapidly down cobblestone streets in a shower of sparks. Fire insurance companies pressured cities to keep trying newer models. Whereas hand pumpers required brawn and bravery, steam engines required operators with the technical know-how to run and maintain complicated machinery. Thus, the switch to steam led to the creation of more professional fire departments, since a salaried engineer needed to be on duty to handle the coal- and wood-stoked machines. For the Boston fire department, the years between 1860 to 1874 were a time of transition from a largely volunteer force to a largely salaried one. The department was soon headed by a "chief engineer" (the term "fire chief" would come decades later), and he oversaw a board of assistant engineers.

The switch to horse power meant other innovations as well. Harnesses were suspended over horses in their stalls and dropped within seconds after a fire alarm, allowing firemen to hook up engines and roll out in less than a minute. Faster speeds meant steam engines were increasingly designed to carry men as well as equipment. The days of "running with the machine" in Boston were at an end.* Within a decade, not only were horses well trained and as responsive to an alarm as any man, but they were lavished with affection and viewed by firefighters as equals in the battle against the flames. A fireman quoted in an article in a *Firemen's Standard* on "famous Boston Fire Department horses" said his favorite horse "could do anything but talk." The king of Boston's fire horses was Fatty, "the oldest veteran and wisest of all the Department's horses." If Fatty could talk, "any fireman will

---

*With one exception. See Chapter 5.

tell you that he could explain how best to control a fire as well as the chief himself." Just as a grateful community attributed sterling qualities to firefighters, firefighters saw those same qualities in their horses: loyalty, spirit, and bravery.

Not everyone, however, adjusted to the transition from manpower to steam power. The Thayer Company gave up in 1860, after making about 100 engines. The switch also proved ruinous for the Hunneman Company. After founder William C. Hunneman retired and his son William Jr. died, another son, Samuel Hewes Hunneman, insisted on keeping the family business together. In 1846 he managed to convince his brother Joseph to leave his New York dry goods business and return to Boston to form the partnership of Hunneman and Company. Samuel died in 1869, and his son, John, ran the company with his uncle. Joseph Hunneman was, however, one of those unfortunate men who think that they can block progress. "There was no greater opponent of the steam fire engine in its early days than [Joseph]Hunneman and not until it was forced upon him did the company consent to manufacture a steam operated machine," Tufts noted sadly. The first Hunneman steam engine rolled out of the shop on May 26, 1866, six years after Boston had converted entirely to steam engines. The Hunneman steamer made up for tardiness with magnificence; it was reputed to be the handsomest engine in New England, if not the world, according to Tufts. With its 265 copper smoke tubes and straight frame engine, "Somerville No. 1" tipped the scale at about 6,500 pounds fully equipped. The company continued to make both hand pumpers and the "newfangled" steamers. Small towns continued to buy the hand pumpers for the next decade, and many of Hunneman's machines remained in service into the twentieth century. Boston, or communities that were later annexed by Boston, purchased eight Hunneman steamers over the next few years. In the late 1860s the city returned to the practice of giving companies a name as well as a number, and companies such as Mazeppa 1, S. R. Spinney 2, Eagle 3, Eclipse 6, and Maverick 9 retained the names of their beloved old Hunneman hand pumpers, although they had converted to steamers.

The Hunneman Company found change unsettling. With its stubborn attention to detail, the company simply could not compete with companies that also built locomotive engines, such as the Amoskeag Company, because these companies could produce the new steam fire machines more efficiently. In 1883, after producing twenty-nine steam fire engines and 716 hand-pump engines, the company went out of business. Joseph Hunneman passed away four years later.

Steamers, too, would see their time come and go. Although by 1909 Boston had two self-propelling steam engines, these would prove no match for the internal combustion engine destined to transform transportation in twentieth-century America. The first gasoline-powered fire engine was purchased in 1910, and over the next thirteen years, horse-drawn equipment was phased out. In 1923, when the last horses were retired from Engine 29 and Ladder 24, the transition to the gasoline-powered era was complete. Some steamers were converted to gasoline, but many ended up rusting in back of stations or in junkyards, alongside the hulks of decaying hand pumpers.

As equipment evolved, fire companies continued to be organized around slightly different functions. Engine companies handled the machines (first hand pumped, then steam driven, then gasoline powered) that were used to pour water on a fire. After strong hose was developed, hose companies were formed to carry reels of hose or piles of folded hose that were too heavy for hand-drawn engines. Ladder companies—first called hook-and-ladder companies—transported ladders and other heavy equipment. As powerful gasoline fire engines emerged, hose companies were disbanded. While firefighters continued to take loving care of their engines, function became more important than form and ornate decorations followed the fire horse into history.*

---

*Today ladder and engine companies have slightly different mandates. Ladder companies provide access to buildings and engine companies provide hose lines, but members of both companies are trained in rescue and in confining and extinguishing fires. Other specialized companies have been added over the years, including rescue companies and marine, tower, and hazardous material units.

Picture a summer day in a field north of Boston. Dotting the field are a dozen hand-pumped fire engines, restored as closely as possible to their nineteenth-century design. In modern-day musters, teams of fire buffs and retired firefighters grab the handles (still dubbed "brakes") of "their" machines and pump—each striving as hard as he (or she!) can to send a stream farther than any other machine and win glory for the team.

Hundreds of hand pumpers (as well as steamers and outmoded gas-powered fire engines) have been saved from the junk pile by admirers who see the history of firefighting written into their mechanics. Many of the old machines have been restored to working condition and are on display in the nation's more than 300 fire museums; some are in private collections. At least 180 Hunnemans are still in existence. J. Richard Hunneman Jr., the great-great-great-grandson of the company founder, has a number of them, which he brings out for parades and special occasions. Moreover, some of the old pumpers are still in hot competition at musters. Through organizations like the New England States Veteran Fireman's League, created in 1891, men (and now some women) can test their mettle on machines made by companies like Hunneman, Jeffers, Leslie, and Button & Blake, competing every bit as fiercely as their nineteenth-century counterparts. During the summer, the machines—sporting names like Hancock No. 128, Governor Bradstreet, and Okommakamesit—are prepped for competition. Gleaming with fresh paint, their brass and wood polished to a high shine, they are pulled into position looking as dressed up as a fire engine can be.

4

STRIKE THE ALARM

The Nerves of Boston's Fire Alarm System

Like the rest of the nation, Boston doctor William Francis Channing was enthralled by news of Samuel Morse's first telegraph message on May 24, 1844, with its evocative question: "What hath God wrought?" What an amazing invention, he thought. Imagine, communication that flew faster than the fastest horse, faster than a bird. Drawing on his medical knowledge, he pictured the electromagnetic telegraph functioning as the nervous system of the nation, linking cities and countries. As the young doctor, himself an amateur inventor, pondered the implications, he wondered if "telegraphy" could help the city of Boston with one of its most tenacious foes: fire. Could the city build a "municipal" telegraph system that could bring firefighters immediately to the exact location of a fire—just as the touch of a finger on a hot coal signaled a quick response from the brain?

Eight years later, aided by the mechanical genius of fellow inventor, Moses Farmer, Channing saw his vision fulfilled: the pair had built the first municipal fire alarm system. It would revolutionize the way fire companies around the world responded to fire. The combined talents of Channing and Farmer—who started as the best of friends and ended as rivals—gave firefighters one of the most potent weapons in their arsenal against the flames: the fire alarm box.

In a world of cell phones and instant 911 connections, those little red boxes in office buildings and schools and on street corners blend into the urban background. We may not notice them—until we need them. Their unassuming ubiquity masks a deeper purpose; even today those numbered boxes are the heart of the Boston Fire Department's rapid response system.

From colonial days on, Bostonians knew that the best defense against a fire was getting to the scene as quickly as possible. People shook wooden rattles or simply called for help, something referred to as "hallooing fire." Citizens—and later volunteer firefighters—would respond by grabbing axes, buckets, ladders, and other gear and racing in the direction of the hallooing. A fire foreman leading the charge might use a speaking trumpet to bark out directions; eventually trumpets became the very symbol of firefighting and a silver trumpet was often a fitting reward for a deserving firefighter. Even today the number of tiny trumpets on a badge signifies a firefighter's rank.

But a trumpet's call carries only so far. Church bells were also rung to call in firefighters. Eventually the city was divided into districts so that when a fire was discovered, the number of that district could be rung on church bells. By the 1840s, however, the growing city was in dire need of a better way of pinpointing the location of a fire.

William Francis Channing

Moses Gerrish Farmer

Almost all cities, except Boston, *have felt the necessity of maturing, to some extent, signals indicating the existence and direction of a fire. In this city, there has been great negligence in this respect and the result is that our engines are sometimes obliged to run wildly about the streets and return home without reaching the place to which they should have been directed.*

That was the conclusion of a writer identified only as "C" in a June 3, 1845, article in the *Boston Advertiser*. "C" had a solution: "By a very simple application of the Electro-Magnetic Telegraph, these evils may be avoided and the means of giving immediate and precise information throughout the city on any alarm."

"C" was William Francis Channing, the son of the city's most illustrious minister and one of New England's loftiest thinkers. As a boy, Channing fought bitterly with his father over his desire to enter the world of mechanics and science rather than study philosophy and religion. He was eventually able to convince his father of the significance of geology and electricity, and his father was able to convince him that the highest goal in life should be to serve mankind. Thus two impulses drove Channing throughout his life: he wanted to make things work and he wanted to make things right. Those combined drives incubated his desire to create a better way to alert the city to the threat of fire.

Channing was born on February 22, 1820, into one of the most influential and prominent families in New England. His maternal great-grandfather, William Ellery, was one of the Sons of Liberty and a signer of the Declaration of Independence. One uncle was a prominent doctor who became dean of Harvard Medical School, another uncle was a Harvard professor and influential editor of the *North American Review*.

Other Channing family connections—a poet cousin who married the sister of transcendentalist Margaret Fuller and befriended Henry David Thoreau and another cousin who participated in Bronson Alcott's Brook Farm communal living experiment—brought William into close contact with nearly every prominent

Bostonian of his day, a time when the city was celebrated as the "Athens of America." In addition, William and his sister were tutored by the formidable Elizabeth Palmer Peabody. One of the three famous Peabody sisters of Salem, Elizabeth was a pioneering teacher, a transcendentalist, the nation's first female publisher, and the founder of the U.S. kindergarten system. (Another Peabody sister married writer Nathaniel Hawthorne; the third married educator Horace Mann.)

Still, none of these illustrious personages was prominent in their day as William's father, Unitarian minister William Ellery Channing. Rejecting Puritan notions of predestination, the older Channing preached a more humanized version of Christianity. As pastor of the Federal Street Church from 1803 until his death in 1842, Channing railed against war and slavery and called for religious liberalism and an American culture and literature freed from the dictates of the Old World. When Channing spoke, the entire city listened; among those moved by his ideas were Ralph Waldo Emerson and Oliver Wendell Holmes Sr. Today his statue stands in Boston's Public Garden with the inscription "He breathed into theology a humane spirit."

So when as a lad William Francis Channing said he would rather invent things than ponder great ideas, his father was horrified. What was the study of mechanics compared to the mind of God? How could mere things, however clever, affect the great social issues of the day? Young William didn't care. He wanted to delve into the principles of nature, not the philosophy of man. Elizabeth Palmer Peabody marveled at the differences between the famous father and his son. William "did not incline to read and Dr. Channing could engage his attention only by reading about the visible universe, voyages of discovery, etc. . . . The history of mechanical discovery and physical experimenting were also his delight," she noted. At age 9 William was fascinated by his toy steam engine. "He invented an improvement which Dr. Channing wrote down with all the child's explanations; and a few years afterwards that very improvement was patented by a later inventor," Peabody wrote.

William eventually studied medicine at the University of Pennsylvania, but although he graduated in 1844 with a medical degree, he never practiced medicine. Instead, he indulged his love of physical science. Like his father, he was passionate about social issues, becoming a fervent abolitionist and a supporter of women's rights. Channing was, however, a new breed of reformer: he believed in the application of technology to solve social ills. In the 1850s he compiled and published a thorough examination of the use of electricity in medicine, declaring, "Electricity is entering year by year more extensively into medical practice and is to become one the most universal and important of the therapeutic agents."

Even as steam engines slowly gained acceptance by Boston's tradition-bound firefighters, innovations such as photography, anesthesia, gaslights, the electromagnetic motor, the sewing machine, vulcanized rubber, and the revolver were transforming other aspects of American life. And dwarfing all of these inventions in psychological impact was Samuel Morse's telegraph, which was heralded as a "most remarkable invention of this most remarkable age." Newspapers proclaimed that the telegraph had "annihilated space and time" and hailed the device as the ultimate demonstration of the nation's genius. Like the hype that would greet the emergence of the Internet more than 150 years later, many proclaimed the invention would transform politics, culture, press, and commerce. Likewise, young doctor Channing was profoundly intrigued by Morse's electromagnetic system, which, he declared, would make possible "a perpetually higher co-operation among men and higher social forms than have hitherto existed."

Channing decided that this new technology could be applied to the goal of protecting Boston citizens from the fires that imperiled lives and livelihoods. Specifically, telegraphy could mirror human behavior. Channing's medical training had taught him to think of the body's nervous system as a series of relays from the brain to the limbs and back again. Fascinated by the exact nature of the impulses between mind and body, he envisioned the electric telegram as a kind of mechanical nervous system. As he wrote in the *American Journal of Science and Arts*: "The Electric Telegraph in its common use . . . is an agency for the transmission of intelligence or impression to a distance. In this its functions are analogous to the sensitive nerves of the animal system."

In June 1845, he published a short article in the *Boston Advertiser* on "Morse's Telegraph for Fire Alarms." He described a central location linked by double wires to fire stations and fire bells throughout the city. Telegraph keys would allow fire stations to communicate with the central office and with other fire stations to coordinate firefighting. Most important, "the agent would be enabled by pressing a single key with his finger at certain intervals to ring out an alarm, defining the position of the fire, simultaneously on every church bell in the city." Channing began to pester Boston officials about his idea. His lobbying so convinced Boston Mayor Josiah Quincy that the mayor called for a fire alarm system in his 1848 inaugural address. Channing's ideas were sound, but it would take another man's genius to make the system work.

Moses Gerrish Farmer was born on February 2, 1820, just a few weeks before Channing. Like Channing, he was a descendant of some of the early seventeenth-century English settlers in New England, but his father, a farmer and prosperous lumber merchant, had a greater interest in the rewards of this life than in those of the next. Moses attended school irregularly and ended up dropping out of Dartmouth College when he became ill with typhoid fever. He made his living as a schoolteacher.

Despite the gaps in his formal education, Farmer was a technological genius. His obsession with mechanics, engineering, and electricity was as fierce as Channing's. He managed to devise a machine that would print paper window shades more efficiently and at far less than the cost of linen. The venture brought him money, allowing Farmer to delve into groundbreaking discoveries in electricity and electromagnetic telegraphy systems. He joined the handful of inventors attempting to find practical uses for breakthroughs in electromagnetism. In the mid-1840s, he created an electric passenger train, which he exhibited with great fanfare in city halls in Maine and New Hampshire. The train was only big enough to carry children, but it was a huge hit. In 1848 he became operator of the telegraph office in Salem, Massachusetts, and opened telegraph offices along this line until 1851.

Farmer was also trying to find a way to create a fire alarm system; he was coming at the problem from a different direction than Channing. In 1847 and 1848 he devised a machine for striking fire alarms on church bells through electromagnetic impulses. He demonstrated the invention to Boston officials, who, though somewhat intrigued, were not interested in laying out money for a project. Still, his interest in improving the fire alarm led to an introduction to Channing, who was promoting his municipal telegraph concept. The two met sometime in 1851 and began to work out details of a municipal fire alarm telegraph system. Both possessed keen minds and extraordinary vision; Channing provided an overall perspective, Farmer was skilled at details. They began a lively correspondence, bouncing ideas and propositions off each other. Their salutations changed from "Dear Sir" to "'My Dear Friend" in letters filled with ideas, sketches, formulas, and diagrams.

In March 1851, Channing made a lengthy and extremely detailed presentation of "the Application of the Electric Telegraph to signalizing Alarms of Fire" to Boston city officials. His plan called for a series of districts, each with a distinct number and a system of double wires linking signal stations to a central office. People would report fires by cranking a handle in the signal box; a notched code wheel would break or complete an electrical circuit, indicating its location by a series of dots and dashes. After verifying the box number, the central office would send out a telegraph signal that would trigger the fire bells, which would chime or "strike"—in firefighting parlance—the number of the district (ranging from one to seven), followed by the number of the box (which ranged from one to ten). For example, to indicate district 2, fire box 5, the bells would strike twice, then pause, then strike five times, then pause and then repeat the sequence.

Channing also noted: "It is obvious that the Chief Engineer might also establish certain signals by which any part of the fire department might be directed either to proceed to a fire or to turn back." He even calculated the length of the required wires and the total cost of the project, right down to the penny: $7,959.60. His arguments were so persuasive that the city voted $10,000 to establish the system on an experimental basis; Morse was paid $800 for the use of his patents.*

Farmer took charge of building the system in September 1851. Using forty-nine miles of wire, he installed forty signal boxes on three signal circuits and nineteen alarm bells on three alarm circuits. The alarm bells would be triggered by his striking machines, some with twenty-five-pound hammers. Telegraph lines were strung from buildings or poles; twofold wires were used to make interruption of the circuits practically impossible. By December wires were strung across the city "like spider webs over a cow pasture on a June morning," reported the daily *Commonwealth* newspaper. "This arrangement is one of the wonders of the age, involving in it one of the deepest mysteries of the universe and the more it is studied, the more mysterious it becomes." Fire boxes, then painted black, were attached to buildings; a central alarm office was set up in the City Building in Court Square. Firefighters called to a fire could tap out calls for more assistance on the telegraph keys in any nearby fire box using American Morse Code. Farmer, the *Commonwealth* reported, "invented much as he went along."

When the system was completed, instructions were posted inside the boxes—and they were far from simple, though Channing told city officials, "The act is so simple that it might be performed by a child." A person calling in a fire had to: "Turn the crank ten or fifteen times slowly and wait. If the signal is heard at the central office, you will know it by the ticking of the district number in the box immediately after. After the sign is thus acknowledged from the central office, you will hear the bells

---

*New York City's fire department also began using a more primitive telegraph system in 1851.

strike the number of the district three times." By April 28, 1852, the system was in place. The next day the first alarm was received at 8:25 P.M. from signal station 7 in district 1. Using a box at a church on the corner of Cooper and Endicott Streets, a citizen attempted to call in a fire at the corner of Charlestown (now North Washington) and Causeway Streets. But the man cranked the machine too fast for the signal to go through, so he ran to the central office to make sure officials knew of the fire. The log reads: "'J. H. Goodale turns the crank like lightning so it could not be read; brought alarm to office." (Quickly, fire box instructions were changed to require twenty-five turns.)

Volunteer firefighters first viewed the system with the same disdain that they directed against the new steam-powered fire engines. "Its utility was disbelieved in by a great many people for a long time, on account of its imperfect manner of working," Farmer later said. Indeed, many technical bugs remained to be worked out. Farmer, who became the system's first superintendent, adjusted the alarm bell apparatus, the hand cranks, and the code wheels. Because fire officials feared false alarms, the boxes were kept locked. Only police and a responsible citizen living nearby had keys. (Keys were not always entrusted to firefighters, either.) The name and location of the citizen were posted on the box, but still anyone who noticed a fire had to search frantically for the person with the right key, often causing a delay. Nonetheless, within the year the system proved its effectiveness, and firefighters, who had been among the fiercest opponents, became among its most fervent supporters.

Channing and Farmer jointly applied for a patent, and patent number 17355 was issued on May 19, 1857. They received a second patent for improvements in March of 1859. Both men, now partners, were determined to sell the new "American Fire Alarm Telegraph" to the rest of the country. Both had received some compensation for the invention, but they were eager to obtain remuneration for years of underpaid work. Farmer, for example, had been persuaded to build the system for $100 a month, a salary he considered grossly inadequate for a man of his abilities.

But he, like Channing, believed the system would demonstrate its usefulness to tightfisted city officials. In 1852 Farmer unsuccessfully pitched the system in Philadelphia and New York. Channing "made every effort in my power by correspondence and personal communications with well-known contractors to introduce the system." He also penned scientific articles and gave lectures, but by October 1855, when he gave a lecture on the system at the Smithsonian Institution in Washington, D.C., he had borrowed money to keep the business afloat and was $5,000 in debt. Despite Farmer's skill and Channing's vision, it took a wily southerner to sell the American Fire Alarm Telegraph to the rest of the country.

Sitting spellbound in the audience at the Smithsonian was a telegraph hobbyist named John Nelson Gamewell. He listened intently as Channing declared, "The first ten minutes in directing the alarm is worth hours afterward." Crediting Farmer, "the ablest and most ingenious telegraphic engineer in the country," with building the system, Channing described in detail how the alarm system operated. He proudly asserted that the loss from fire in

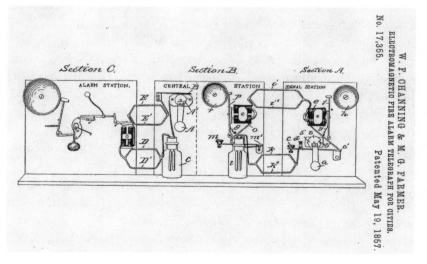

Illustration of the "fire alarm telegraph" included in the patent for the device

Boston in 1854 was only $150,772, "or less than one dollar for every inhabitant; a loss which, for its small amount in so compact and wealthy a city, cannot be paralleled in America." With a flourish he concluded: "But more than this, it is a higher system of municipal organization than any which has heretofore been proposed or adopted. In it, the New World has taken a step in the forms of Civilization in advance of the Old." Gamewell was thrilled, not about the social significance of the system but about the vast amounts of money that could be made from it.

Gamewell was born in 1823 in South Carolina, the son of Frank Asbury Gamewell, a pioneering Methodist minister. His two brothers became preachers, but John was fascinated by electricity and became a postmaster and telegraph operator for Camden, South Carolina. A man of little means (his father had died when he was 4), Gamewell had a personality of great persuasion. Drawing on the goodwill of friends, he raised $30,000 and bought patent rights from Farmer and Channing for marketing the system in the South; soon after, he bought the rights for the rest of country. Gamewell proved to be a more successful salesman than Farmer or Channing. Philadelphia was his first customer and installed a system by April 19, 1856. Orders came in from St. Louis, Baltimore, New Orleans, and Charleston, South Carolina. In 1861, just as Gamewell had begun to build a fire alarm empire, the Civil War erupted. As a Confederate, Gamewell wound up on the losing side, and the U.S. government confiscated his patents. John Kennard, a Boston fire alarm official, was quickly dispatched to buy them back for Boston. He was willing to pay as much as $20,000; he got them for $80. Forming his own company, Kennard installed the Charleston, South Carolina, fire alarm system by May 1861 despite the ongoing war.

The enterprising Gamewell, while financially destitute, didn't stay down for long. After the war, he joined Kennard's company and the new Gamewell, Kennard & Co. set up a manufacturing facility in Upper Newton Falls in Massachusetts. Kennard later dropped out to serve as superintendent of the Boston fire alarm system from 1867 to 1880, and the firm became the Gamewell

"Original signal box," 1851

Company. Gamewell remained determined to sell his system to every American city. He and his employees, who included engineering wizards Moses Crane, Edwin Rogers, and James M. Gardiner, continued to improve the system and attacked with gusto any other company that dared to get into the business. Said one Gamewell brochure of its competitors: "Careful examination has always shown their apparatus to be largely made up of devices infringing on our patents, yet crude, fragmentary and very unreliable in operations."* In the post–Civil War boom, cities began to adopt the new system; by 1871 about 50 cities and towns had installed fire alarm systems, most of them from Gamewell. By 1884 Gamewell systems were installed more than 150 cities and towns. When Gamewell died in 1896, at age 73, his name had become—and still is—synonymous with fire alarm systems.

Boston continued to improve its fire alarm system over the years, first under Farmer, then under Joseph Stearns, who succeeded him as superintendent in 1853, and later under Kennard. In 1864, as Boston began to annex surrounding communities, the

---

*In 1884 Gamewell also warned against relying on the telephone: "one of the most sensitive of instruments, easily affected by disturbing causes and therefore entirely unreliable for fire telegraphs."

system of ringing out district numbers was changed; instead, boxes were numbered from one to seventy-three, meaning that now bells would ring out the box location, not the district number. By 1869 the boxes were modified to allow the pull of a handle to set the code wheel in motion, replacing the need to crank a handle. (At first, however, people had a tendency to pull the handle repeatedly or pull it down and not let go.) On May 21, 1881, the city directed that all boxes be painted red. Later that year the city installed boxes that required no keys; however, a small alarm sounded when the box was opened, to discourage false alarms. In 1928 the door alarm system was discontinued because people too often thought they had sounded an alarm just by opening the door.

As firehouses were increasingly staffed with professionals rather than volunteers, the need for public fire alarms faded. The use of church bells was discontinued by 1909, and the last public fire bell was removed from Fanueil Hall on May 25, 1929. Over the years, the fire department developed procedures that dictated which fire companies and apparatus responded to particular alarms. A second alarm brought more help, a third alarm even more, and so on. The order was listed on running cards, so called because firefighters used to run alongside the machines with the information. The fire alarm office was moved to the newly built Boston City Hall on School Street in 1865; in 1895 it was moved to department headquarters on Bristol Street. In 1925 the office was relocated to an imposing structure on park land in the Fenway area of Boston, where it remains today. Two-way radios were introduced in 1923—the first to be used in a fire department in the United States. Initially the radios were used to communicate with fireboats; eventually they were provided to chiefs, and by 1953 every piece of apparatus was equipped with a radio. By the late 1960s portable radios were in operation. Yet the basic concept developed by Channing and Farmer remains intact.

In creating the fire alarm telegraph system, Channing found a bridge between his inclination for science and his father's aspirations for humanity. He simply decided to save lives, while his father sought to save souls. But Farmer, who was by all accounts a greater innovator than Channing, was always more interested in obtaining compensation for his inventions than deriving satisfaction from their impact on the greater good. And he had just cause for feeling slighted because, as the system was adopted around the country both he and Channing realized they had been woefully underpaid. Farmer had been forced to sell his share in the business to Channing in 1867 due to financial setbacks stemming from a financial panic in 1857. In 1871 the pair fought a protracted legal battle to get an extension of their original patent. During a hearing before the Commissioner of Patents, several city fire officials testified to the great value of the system. An Albany, New York, fire commissioner declared that the fire alarm system saved his city a quarter of a million dollars annually; another city official stated that if the system were put into use in all major U.S. cities, the savings would be "at least two million a year." Yet Farmer and Channing testified that they had made only $30,000 from the patent, an amount "entirely inadequate for so valuable an invention." The pair won the extension, and the rights were transferred as previously agreed to the Gamewell company, which had paid the legal expenses.

But Farmer and Moses continued to wrangle with Gamewell over proper compensation, and although they had enjoyed an unusually fruitful partnership, tension now strained their relationship. Farmer became increasingly disgruntled over the adulation that Channing was receiving and the money that Gamewell was making and wanted it known that he, after all, built the system. He wrote increasingly harsh letters to his co-inventor, complaining about his lack of recognition and insisting that "I need all that is due to me" from Gamewell. Channing repeatedly tried to smooth Farmer's pride and save the friendship. In October 1877 he made a last entreaty: "I will repeat what I have said: that I consider you my equal and peer in the creation of the American Fire Alarm Telegraph, that I had always meant that this should be recognized. I shall always take pleasure in helping to make it apparent."

Perhaps Channing's repeated reassurances only rankled Farmer further with their undertones of condescension and noblesse oblige. Channing apparently had fewer financial woes than Farmer and was more concerned with receiving, as he put it, "the satisfaction of contributing in one small particular to the progress of civilization." He had, for example, attempted to make sure Boston would never have to pay for patents on the fire alarm system. On July 1, 1858, he had signed a "covenant" stating that for the "sum of One Dollar," he and his heirs would give Boston the

> *full right and privilege to apply and use in the said Fire Alarm System so as aforesaid established and who in operation in the City of Boston in the county of Suffolk and Commonwealth of Massachusetts any and all improvements in the Electro-Magnetic Telegraph Fire Alarm system or apparatus, which may be hereafter invented by me or for which Letter Patent of the United States may at any time be granted to me, or, being granted to any others, assigned by them to me.**

Farmer went on to a life of invention and thwarted ambition. His output was prodigious: he continued to make improvements on the telegraph system, including development of a synchronous function by which multiple telegraph messages could be sent at the same time. In 1859 he lit the parlor of his Salem home with incandescent electric lamps powered by galvanic batteries—this at a time when homes were lit by candlelight, or at best, gaslights. The incandescent lightbulb, perfected by Thomas Edison in 1879, proved to be more practical for mass consumption, however. Farmer also felt he had come close to inventing the astonishing device that would transform communication. He had been a mentor to young Thomas Watson before Watson assisted Alexander

---

*This covenant was discovered in pieces in the archives of the Bostonian Society, so it is unclear whether the city ever received such a document. Testimony in the patent case indicates that Boston was to have "an unrestricted use of Channing and Farmer's invention in return for expenditures" on the alarm system "experiment."

Fire alarm office, circa 1902

Graham Bell in the invention of the telephone. In his autobiography, Watson described how Farmer dropped by to see the "latest developments" and afterward "we sat down and had a chat over old times."

> *He had tears in his eyes as he told me that, when he first read a description of Bell's telephone and realized what a simple thing it was, he was not able to sleep for a week, he was so mad with himself for not discovering the thing years before. "Watson," he said, "that thing has flaunted itself in my face a dozen times during the last 10 years and every time I was too blind to see it."*

From 1872 to 1881 Farmer was a professor of electrical science at the U.S. naval torpedo station and created the four-function relay for the guidance of torpedoes. He made discoveries in magnetism, alloys, and electric dynamos. Indeed, Farmer has often been credited with laying "the foundation for electrical engineering in the United States," according to the *Twentieth Century Biographical Dictionary of Notable Americans*. Yet today we recall only names like Alexander Graham Bell and Thomas Edison. As Massachusetts Governor William Claflin said of Farmer after his death, "He was deserving of more honor than he ever received," and a fellow electrician lamented that Farmer "invented not wisely but too well." He died on May 25, 1893, in Chicago, where he had gone against his doctor's advice to set up a display of his inventions at the World's Fair. His obituary highlighted his work on the fire alarm system, crediting him with conceiving it in 1847–48 and building it in 1851–52. Channing is never mentioned.

Channing's wide-ranging interests took him in new directions. He remained active in various social movements, and while living in Providence, Rhode Island, he conducted experiments designed to improve upon Alexander Graham Bell's telephone innovations. Seeking to solve the problem of long shipping routes around the tip of South America, Channing proposed the creation of a "ship railway" that would carry ships across the Panama isthmus. In March 1880 he presented his designs to a committee of the U.S. House of Representatives. The U.S. government, however, chose to go with a more conventional solution and built the Panama Canal. Even in the last years of his life, he remained fascinated with science and its application to religion. Just a few months before his death at age 81, he was still trying to connect his inclinations with his father's desires by planning to write an article that would reconcile the "immanence of God" with the scientific principles of the "ether" of space.

Channing died on March 20, 1901, a few months into the twentieth century. His obituary, like that of Moses Farmer, chiefly focused on his invention of the fire alarm, but unlike that of his co-inventor, it fully acknowledged that Channing had had "most valuable assistance from Moses Farmer." Despite their rivalry, the name of Channing and Farmer will remain forever linked in fire history.

POSTSCRIPT

Boston's fire alarm system has been enhanced by computers, sophisticated backup batteries, improved electronics, and the addition of radio communication. Yet its operating principles remain essentially the same as those envisioned by Channing and Farmer. The Gamewell Company, based in Ashland, Massachusetts, continues to make fire alarm and security systems, although the company has undergone several ownership changes.

About 1,350 fire boxes—virtually all of them still made by Gamewell—dot Boston's streets; another 1,300 master boxes are in public and private buildings. Each one still has a spring-wound coded wheel inside, which is triggered when a handle is pulled. The boxes are connected on a "closed circuit" system to the fire alarm headquarters, now located in the Fenway area of Boston.

A classic 1896 Gamewell box stands outside the National Fire Protection Association headquarters in Quincy, Massachusetts

In 1988 computer terminals, which could decode the location of boxes when alarms are pulled, were installed. In 1994 a sophisticated computer-aided dispatch system was added.

When a fire is called in by telephone, the department's computer system locates the nearest box. A dispatcher transmits the box number over the wire circuits, which strikes the number of the box in Boston fire stations. The dispatcher also broadcasts the address by radio, followed by the box number, designated in a series of digital tones. Thus Box 1591 would be signaled by one, five, nine, and one "beeps." Firefighters assigned to respond to that particular alarm will, when needed, radio fire alarm headquarters to increase the number of alarms and send more companies. This is signaled by adding a prefix to the box signal: a second alarm, for example, is signaled by two beeps followed by the box number. After the "all out"—the code 22-22—is sounded, a fire chief or captain will open the box and rewind the coded wheel. The department also has "phantom boxes" or designated numbers for corners that no longer have boxes. Some cities have eliminated fire alarm boxes, but Boston officials consider the system essential. Over the years Boston has annexed many adjoining communities and thus has duplicate street names—there are multiple Washington and School Streets, for example. Box numbers, however, are unique. Signal boxes are "multilingual," which—given Boston's large immigrant and student population—eliminates problems of accents or confusion over directions. Firefighters often know from memory the location of every box in their district.

In more than 150 years of operation, the fire alarm system has never once broken down.

5

THE GREAT FIRE OF 1872

A Disaster Foretold

Saturday, November 9, 1872, was the kind of glorious fall day found only in New England. With temperatures unexpectedly mild and the air calm and clear, John Damrell had every reason to hope he could finally spend a quiet evening in his Temple Street home on Boston's Beacon Hill.

For two weeks, the chief engineer of Boston's fire department had been troubled by visions of last year's great Chicago fire. In October, a severe outbreak of horse distemper, something called an "epizootic," had mercilessly spread among the region's horses, killing some and sickening all. With their specially trained animals too weak to work, firefighters had to haul their engines and equipment by hand, crippling their ability to get to a fire promptly. Fearing Boston's rapid growth had created the same conditions that turned the Chicago blaze into an inferno, Damrell and his assistant engineers quickly devised an elaborate contingency plan. It had worked surprisingly well, and as a spectacular sunset brought November 9 to a close, the veteran fireman was sure the crisis was nearly over.

Just before half-past seven in the evening, he heard the fire bells ring. Five blows. Pause. Then two blows. Box 52. That was the intersection of Summer, Lincoln, and Bedford Streets, the heart of the city's wholesale district, probably the worst place in the city for a fire. He threw on some clothes, grabbed his white chief's hat, and dashed out into the warm night air. As he sprinted around the corner of Park and Beacon Streets in the direction of a strange glow, he heard a second alarm. And when he reached the corner of Summer and Kingston Streets, he saw a sight he had never seen in his twenty-five years of fighting fires.

A four-story granite building was now one vast furnace, as if the earth itself had broken open to reveal its molten core. Ten- to twenty-pound chunks of flaming granite were bursting from the blaze as if shot by a cannon. Around him, sparks flew as thick as flakes in a snowstorm; above him, hoop skirts, transformed into rings of fire, sailed on the thermal drafts. The heat was so intense that he could feel his skin burning at fifty feet away. Flames were also ripping through a second building on Kingston Street and

Mayor William Gaston, Fire Chief John Stanhope Damrell, and Police Chief E. H. Savage, pictured in the frontispiece of Russell Conwell's History of the Great Fire in Boston, *published shortly after the fire in 1873*

another on Summer Street. To his great relief, he saw that men from three of his fire companies, Engine Company Nos. 4 and 7 and Hose Company No. 2, had already taken up positions as close as they could to the ferocious heat.

"For God's sake, hold the corner," Damrell barked, as he tried to get a fix on the trajectory of the flames, which were spreading faster than he had ever seen a fire spread. His worst nightmare had flared into reality; the blaze was fast becoming an inferno that would rival the Chicago fire. In Chicago, Mrs. O'Leary's cow got the blame. In Boston, accusing eyes would focus on Chief Engineer Damrell.

Drawings and photographs of John Stanhope Damrell show a man with intelligent, lively eyes, a receding hairline, a mustache, and then-common muttonchop whiskers. A master builder by trade and an innovator by nature, John Damrell was the kind of fire chief who believed it would be better to answer four false alarms than allow any delay in reaching a real fire. Born in Boston's North End on June 28, 1828, and orphaned at an early age, he became a carpenter's apprentice and by 1856 had launched a building partnership. An avid volunteer firefighter, he joined Hero Engine Company No. 6 when he was barely out of his teens—both his father and brothers had served as firefighters, too. He was popular with his fellow firefighters, who admired his technical skills as well as his willingness to roll up his sleeves and jump into action. In 1858 he was appointed an assistant engineer, and he was continually reappointed thereafter. Damrell had a keen sense of civic responsibility; he had once been elected to Boston's Common Council.* Then, in 1866, he was elected chief engineer by the city council in a fiercely contested race; the newly organized Engine Company No. 11 of East Boston was christened the "'John S. Damrell" in his honor.

Damrell approached firefighting with curiosity and gusto. After the devastating 1871 fire in Chicago that killed 300, left 90,000

---

*The Boston City Council then comprised two bodies, the Common Council and the Board of Aldermen.

homeless, and destroyed more than 2,000 acres and 18,000 buildings, he traveled to that city to learn all he could about its firefighting efforts. He had a long discussion with General Philip Henry Sheridan, a Civil War veteran and a hero of the Chicago blaze, and came back convinced that Boston stood on the brink of calamity.

Unlike firefighters of the past, Damrell welcomed new technology. He commissioned the building of the city's first fireboat, eventually deployed in 1873. He successfully lobbied the state government to give the fire department the right to make building inspections. For five years he badgered the Boston City Council to improve water systems and update fire hydrant couplings. He was particularly concerned about the tall buildings, narrow streets, and corroded water mains in the city's burgeoning commercial district. Formerly a residential area where Ben Franklin and Daniel Webster once lived, the blocks from Washington and State Streets to the harbor's edge had been transformed in the post–Civil War boom. The area now housed many of the city's wholesale leather and clothing businesses as well as small factories, banks, retail stores, and newspaper offices. But growth had outpaced city services. To increase water pressure to ensure a steady flow for fire hoses, Damrell wanted to replace the area's six-inch water mains—reduced to four or five inches by corrosion and rust—with pipes at least eight inches in diameter.

But city councilors brushed him off, saying, "Do not try to magnify the wants of your department or your office so much." As one historian has noted, "The hard-headed Yankee gentlemen who sat in the city council were not far seeing and they abhorred extravagance."

Damrell had no desire to be extravagant. Rather, he was what could be considered Boston's first modern, professional fire chief. He was more anxious to prevent fires than to devise clever ways to get there first. As a hands-on manager, he believed in leading by example through working the fire lines alongside his men. This was one of his greatest strengths, but in November 1872, it became his greatest weakness.

At that time Damrell oversaw a 475-man department. About 385 were "call" men who came to the fire only as needed; the other 90 permanent members were mostly engaged in driving and operating equipment. The firefighters were divided among twenty-one steam fire engine companies, seven hook-and-ladder companies, ten hose companies, and three chemical extinguisher wagons (which mixed soda ash and other chemicals to produce water pressure in a small tank), spread from Boston proper to South Boston, East Boston, Roxbury, and Dorchester. Damrell answered to the Committee on the Fire Department, a body composed of eight city council members appointed by the mayor. A board of fire engineers, somewhat analogous to assistant district chiefs today, were appointed annually by the city council and answered to Damrell.

In the fall of 1872 Damrell and his engineers were staring calamity in the face. The virulent form of horse distemper, which had originated in Canada, reached the East Coast. While not always fatal, the animals quickly became too sick to work, often dropping in their tracks. Transportation came to a standstill. Train depots filled with undeliverable goods, men were forced to pull horsecars and garbage carts, and many seized the opportunity to charge exorbitant fees for hauling wagons. The modern equivalent of this "epizootic" would be a complete and total gasoline shortage.

When the department's horses succumbed, Damrell acted quickly. On a rainy Saturday, October 26, Damrell and his assistant engineers met at City Hall on School Street to work out a contingency plan. First, they decided against trying to obtain other horses—even if they could—because fire horses were specially trained not to run from smoke and fire. Instead, men would return to the practice of "running with the machines." Damrell and his assistants decided to hire 500 extra on-call men, paying them $1 for showing up to a fire and 25 cents an hour thereafter. Damrell also worried about leaving parts of the city unprotected if too many firefighters rushed to a blaze. Normally, the number of engines responding to a first alarm varied; more engines were sent to potentially more dangerous sections of the city. The board decided that, temporarily, only two engines would respond to first alarms. If a second alarm were struck, all assigned companies would then respond. Due to a rash of false alarms when first installed, fire alarm boxes were then locked, and only policemen and a "responsible" person in the neighborhood had keys. So as an extra precaution the engineers decided to instruct police to send out a second alarm if a fire was seen above the third floor, where water could be reached only by steamers. Ordinarily, police would wait until an engineer arrived to send a second alarm. Damrell left the meeting, relieved that the city was protected. The next two weeks seemed to prove him right; nine fires were successfully fought, and four days had passed without any alarms at all. He and his men had, however, made "what proved to be a serious blunder, none the less deplorable because [it was] the work of brave and conscientious men," Murdoch wrote.

On November 9, many Bostonians strolled through the city's streets, enjoying the unseasonably warm Saturday night. Crowds hovered at the doors of the brightly lit theaters; a citywide gas system installed in 1828 had eliminated reliance on oil lamps and candles. The chief attraction was E. A. Sothern in *David Garrick* at the Globe Theatre. Gilman Joslin Jr., a resident of Charles Street, was at a nearby YMCA for a lecture. Sarah Putnam, a portrait painter and member of a wealthy Brahmin family, had seen *David Garrick* the previous evening and so was spending a quiet evening at home. The talk was of General Grant's apparent triumph over Horace Greeley in the presidential race and of a gruesome Cambridge murder on election night; parts of the body of Elijah Ellis were found floating in the Charles River.

The city's commercial district was quiet and nearly deserted. William Blaney, supervisor of the basement boiler in a four-story granite edifice at the corner of Kingston and Summer Streets left at about 5:20 P.M. Everything, he later recalled, seemed in good order. The six-year-old building housed the wholesale dry goods store of Messrs. Tebbits, Baldwin, and Davis in the basement and

Where the fire started: the corner of Kingston and Summer Streets

first floor. Damon, Temple, and Company, wholesale vendors of hosiery, gloves, laces, and small wares took up the second and third floors. Alexander K. Young and Company, which manufactured hoop skirts, occupied the upper floors. The structure was topped by a wooden "mansard roof," a popular steep-slated style that originated in France and was widely used in Boston. A small steam engine in the basement ran the elevator. The building was jammed with merchandise for winter stock, including rolls of cloth and bundles of hosiery, gloves, and laces.

No one ever determined how the fire started. Sometime before 7 P.M. a spark ignited flammable material in the basement of a building at the corner of Kingston and Summer Streets. "Some spark snapping outward, or some stray coal undiscovered, which escaped from the furnace during the process of raking, or overheated surfaces which came in contact with combustible material may have caused the fearful destruction," wrote Russell H. Conwell in his history of the fire. "No human eye was there to note the little spark, the diminutive flame and the tiny stream of smoke that could so easily have been smothered with the foot, or extinguished with a cup of water." (Like many of the fire's chroniclers, Conwell relished extravagant speculation.)

The fire reached the wooden elevator shaft, which quickly ignited, sending flames rushing toward the magnificent mansard roof. Just after 7 P.M. Charlestown police officers on duty at the Prison Point drawbridge saw a glow in the sky. "Must be a fire in Boston," one remarked. Although two miles way, they were the first to spot the blaze. About 7:10 P.M., John Holmes and Augustine Sanderson were walking through the area when they spotted flames that had burned through the building's roof and, more surprisingly, had already attracted a dozen onlookers entertained by the crackling flames. Sanderson started running down Summer Street, "hallooing fire." Yet no one triggered the locked fire alarm, although Sanderson called upon bystanders and even a policeman for help. Police officer John M. Page, investigating commotion among boys on Summer Street, saw the flames. He ran to Box 52, unlocked it, and cranked the alarm; it went out at 7:24 P.M. Immediately he gave a second alarm, which went out at 7:29 P.M. A fire engineer who reached the scene a few minutes later yelled to Officer Page to sound the third alarm; it went out at 7:34 P.M. But a fourth alarm—meant to call every fire company in the city—was not triggered until 7:45 P.M. The fate of the city may have been determined in that delay.

When Damrell heard the alarm, he already knew that Box 52 was what firefighters called a "bad box," because of its location in the commercial district. Normally, six steam engine companies were supposed to respond to this box on the first alarm. The epizootic had changed the rules; only two engines were to respond. By the time Damrell got to the corner of Kingston and Summer Streets, the fire had begun its inexorable march. The few residents of Kingston Street were already hastily gathering possessions for a rapid retreat. Witnesses could hear the fire roaring like a blast furnace, punctuated by the crack of windows shattering. Three steam engine companies took positions at Summer and South Streets, Kingston and Summer Streets, and Kingston and Bedford Streets. (Engine 10 had defied the emergency orders and had responded to the first alarm.) The mansard roofs were proving irresistible to the flames, and although the steam engines were chugging, the streams of water were weak and did not reach above the fourth floor to the fire that was spreading roof to roof. Eight-year-old John Taber helped firefighters drag their hoses; fifty years later he would be chief of the Boston Fire Department.

"Sound the general alarm," Damrell shouted to his assistants. He pulled another assistant aside and told him to send telegrams to all fire departments within fifty miles for help. A member of Engine 4 rushed in to say they could not remain in position because of the heat. A huge piece of granite had split off, and fell, breaking the suction hose; Damrell told him to shift machinery and spray the men as protection. Soon firefighters, civilians, and city officials, begging for his attention, beset Damrell on all sides. Even a 12-year-old lad ran to him and grabbed him by the hand, crying, "My mother and father are in there, please help them." But it was too late. Damrell saw to his horror that it was impossible for him to get into the building.

Trying to get a fix on the fire's perimeter, Damrell ran three blocks to Milk Street, where he smashed in the door of the tallest building with his ax and, lantern in hand, climbed to the sixth floor roof to survey the fire. With a fear close to panic, he saw the disaster unfolding. If the fire spread to South Boston, a place filled with mills and stables, it would "be terrible beyond description." Already he was considering the most terrible weapon in his firefighting arsenal—gunpowder.

In the nineteenth century, gunpowder explosions were an avenue of last resort to halt out-of-control fires. If a blaze leveled buildings in its path, no-burn zones would be created to halt its spread. Explosions had to be precisely set and timed fuses used; the danger to firefighters and bystanders was substantial. A misplaced explosion might even spread the fire it was intended to halt, and afterward there would be complaints and even lawsuits from owners of the blown-up property. General Sheridan had cautioned Damrell about the use of gunpowder, saying press reports of its success in Chicago were exaggerated.

Franklin Street on the night of the fire

All this was on Damrell's mind as the fire illuminated the Boston sky. He also knew that if he didn't use gunpowder, he would be accused of hesitating to take bold steps. Terrible choices lay before him.

By 8 P.M. all twenty-one Boston engine companies were either at or racing to the fire, and companies from Cambridge and Charlestown were also on their way—virtually all were pulling their engines by hand. In just thirty minutes the men of Ladder Company Number 7 sprinted with their equipment four miles from their station on Parish Street in Dorchester into Boston, arriving at 8:15. By 9 P.M. the fire was devouring the large commercial building known as the Beebe block in Winthrop Square. By 10 P.M. flames had spread two to three blocks in three directions to the edge of Franklin and Federal Streets. The mild wind was on the side of the firefighters, but the blaze's intense heat created its own whirlwinds, wailing at thirty to thirty-five miles an hour. Bizarrely, the fire spread *against* the wind with "a malignant relentlessness," as one observer said. By 11 P.M. fiery tongues were licking the wharves along Boston Harbor and the schooner *Louisa Frazer*, which was not moved to safety, was consumed. A tugboat, quickly commandeered as a fireboat, managed to save the bridge into South Boston.

Firefighters would spray the flames as long as they could, often lying in gutters to escape the heat, but the streams from the engines were weak and the men constantly had to regroup and fall back as the narrow streets turned into block-high blow-torches. "The flames ran rapidly up Summer Street, and each street opening on to it became at once a funnel through which the fire poured with inconceivable force," *Harper's Weekly* reported. "Catching at the fatal Mansard-roofs, it went roaring and crackling along the streets, wrapping block after block in flames. The scene was one of dreadful magnificence."

By midnight thousands of Bostonians had turned out to watch their city burn. At about 8 P.M., Sarah Putnam, sitting comfortably in the parlor with her family, was interrupted by a neighbor who came in to tell them of the fire just across the Boston Common.

Sarah and her brother John decided to take a look and, gathering other neighbors as they walked, they arrived at the Common to find it as bright as day. The scene was eerily beautiful, Putnam wrote in her diary. "The trees on the common were lovely, with the branches all illuminated." Transfixed, they watched the flames until 1 A.M., as masses of people carting possessions and goods poured into Boston Common, and "the fire kept bursting out in awful great masses." Putnam also recorded her mother's description: "All Boston was in the streets. The windows of all the houses on Beacon, Arlington, Marlboro, and adjacent glowed as though they were themselves full of fire."

When Gilman Joslin's lecture was interrupted by someone saying a fire was visible, he, like others, paid little heed. "I hadn't the slightest thought, for half an hour, that it would be allowed to spread," he later wrote his brother. He even thought his colleagues were overreacting when they wondered if their stores might be threatened. But he decided to take a look. By 10 P.M. he had forgiven his colleagues. "Heavy granite walls could be seen falling in all directions, and these, with the bright glare, the sharp explosions of buildings, the racket of the steamers . . . produced altogether an effect not to be forgotten," he wrote. In the raging inferno, granite seemed to melt rather than burn. Broken lampposts flared like rockets, fed by gas mains that city officials were unable to turn off. Few observers would ever forget the unholy light that turned Boston's streets as bright as day; a passing flock of ducks, reflecting the light, was first thought to be a meteor shower. The glow in the sky could be seen as far away as New Hampshire. Twenty-one miles away, in South Abington, a scorched $50 bill was picked up. Conwell called it the "war dance of the fire-fiends, with all its hideous concomitants—its snapping, rattling, bellowing, crashing; its steams of hellish flame, and puffs of swarthy smoke, as though the earth had yawned and loosed those weird, traditional denizens of its fiery depths."

Why did the fire spread so quickly? Many blamed the ornate mansard roofs, observing that the roofs caught first and the buildings burned from the top down. A year earlier, Watertown

"City in Flames," Currier & Ives lithograph of the great Boston Fire (top); firefighters try to save the Boston Pilot Building, which had the kind of mansard roof that seemed to contribute to the fire (bottom)

resident Joseph Bird, who had made a personal study of fires, had published an article strongly warning that the mansards were "perfect tinder boxes." Damrell had also warned the fire department committee that mansard roofs helped spread the Chicago fire. Hampered by initial delays, firefighters were further constrained by the narrow streets, the lack of water pressure, and the huge crowds of gawking onlookers. Damrell was confounded by the speed with which the initial building was engulfed. "I don't understand it today," he said months later. "It is a phenomenon which I cannot possibly fathom."

Firefighters from across New England poured into the city. Nearly 1,700 men and 100 companies came from 27 communities, some from as far as New Haven, Connecticut, and Providence, Rhode Island. Self-propelled Hunneman steamers, like the Kearsarge from Portsmouth, New Hampshire, were loaded onto special trains and carried into the city. Out-of-town firefighters, however, often found their hoses' couplings did not match Boston's outdated hydrants, and they wasted precious time adjusting the couplings to fit.

As Saturday turned into Sunday, Boston's strangely bright streets teemed with people: families clutching their children and possessions, merchants with carts trying to haul away merchandise, businessmen carrying accounting books and ledgers. Frantic merchants attempted to bribe firefighters or fellow citizens to save their merchandise by dousing it or carting it away. One newspaper report said as many as 100,000 thronged the streets. Since Boston then had a population of 260,000, this would have been a crowd indeed.

Some onlookers came to gawk and stayed to help. The one triumph of the night came at Hovey's Department Store on Summer Street, where employees and volunteers valiantly soaked rugs and rags, hung them out windows, and stayed inside to stamp out sparks. They had precious little water; only a thin stream trickled from faucets, but employees patiently filled buckets, hauled them up the stairs, and poured the contents onto the carpets. A cheer rang out when firefighters finally managed a

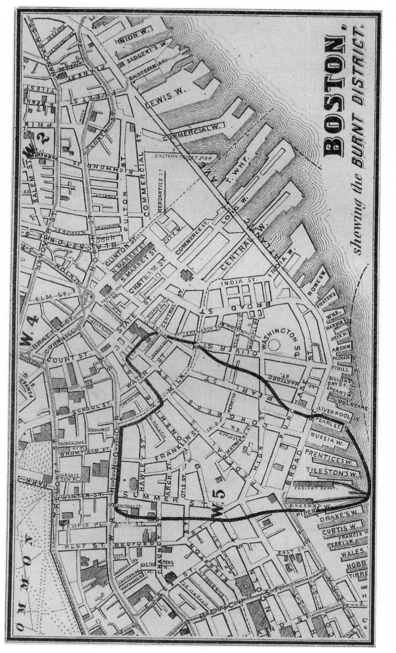

Map of the area affected by the 1872 fire, from Conwell's History of the Great Fire in Boston (1873).

small but steady stream onto the building. Although drenched, scorched, and saturated with smoke, Hovey's survived.

In other parts of the city, however, the fire burned with no one to stop it, neither firefighters nor onlookers. "A building all dark would be a living furnace in five minutes. It was the effect, I suppose, of the tremendous heat making its way through the walls," wrote a young "gentleman" in a lively series of letters on the fire. (The letters and their author, who managed to be in every decisive scene of the fire and who made remarkably factual and amusing observations, was apparently a creation of historian Harold Murdoch in his literary version of today's "docudrama.")

Many prominent citizens raced to the fire with a mix of curiosity and concern. Dr. Oliver Wendell Holmes, the philosopher, poet, and father of a future Supreme Court justice, watched the flames as they crept toward a bank on State Street where he had deposits. "I saw . . . the fire eating its way straight toward my deposits, and millions of others, with them and thought how I

"Into the Jaws of Death": a contemporary engraving from Harper's Weekly

should like to have them wiped out with that red flame that was coming along clearing everything before it," he wrote a friend. "My son Wendell made a remark, which I found quite true, that great walls would tumble and yet one would hear no crash— they came down as if they had fallen on a vast featherbed."

Harvard president Charles Eliot rushed in from Cambridge to look after Harvard's Boston properties; he and other Harvard officials managed to obtain the college's securities from a threatened bank, pile them into a carpetbag, and carry the bag over the Charles River. When the flames came perilously close, officers of the Five Cent Savings Bank on School Street carried $11 million in money and securities to the home of bank president Paul Adams on Charles Street. Legend holds that Adams, shotgun in hand, guarded the assets.

Some merchants, seeing their stores were doomed, decided they'd rather see their stock used than burned; they began giving away shoes, hats, and clothing. This was, Damrell later said, "one of the worst things that were ever done." Soon folks were running through the streets, clutching suits and shoes, and the police could not determine who was a looter and who was a lucky recipient of fire-inspired largesse. About 300 people were detained by police until the mess could be sorted out, and about 70 were eventually convicted; about $80,000 in suspicious goods were stockpiled by police. Damrell himself personally rousted about fifty people from a burning building; "so eager were they to plunder, they never attempted to save property."* Two men inside a burning store on Washington Street (who were attempting to salvage material either for their employer or for themselves) were trapped when a wall collapsed. Hearing cries for help, Boston firefighters William Farry and Daniel Cochran of Ladder Company Number 4, rushed into the store. Another wall collapsed and all were crushed to death.

---

*According to historian Susan Wilson, the Episcopal bishop of Boston, Phillips Brooks, ran into the owner of the elegant jewelry store Shreve, Crump, and Low, who was worried about his stock. Brooks, at six feet four inches and 300 pounds, agreed to fill his oversized coat with jewels and carry them to safety. But the prominent clergyman was promptly detained for looting.

Some observers said firefighters were not above helping themselves to merchandise in abandoned stores. Eliot (whose father,
Samuel, had put down the Broad Street riot as Boston mayor)
later said he overheard firemen complaining that police prevented them from getting their fair share of clothing. "I'm sorry
to say, they were from Cambridge," he haughtily added. Damrell
also "met a fireman with a suit of clothes and I immediately took
him by the throat and saw he was an out of town man and spoke
very strongly to him."

But the vast majority of firefighters fought the blaze with
courage and at huge personal risk. "Never did a body of men
work more heroically or better," Damrell said. Firefighters would
"play" the fire—that is, pour streams on it—until driven back by
the heat. Time after time, streams would run weakly or not at all.
The effect was demoralizing. Assistant engineer John Jacobs was
one of those who worked nearly two days without a break, until
he checked into an infirmary, his eye so blistered by cinders he
could barely see. Jacobs described taking a comrade "by the col-
lar" and "throwing" him up on a roof of a building, where "he
was there like a bulldog and would not leave until he was driven
out." When the steam engines ran short of fuel, firefighters
burned everything they could find, boxes, blinds, fences, coun-
ters, and shutters. The draft was so strong, "we could hardly hold
our hats on," said Charlestown chief engineer William E. Delano.

"What man could do the firemen of Boston did," Conwell
wrote. "They erected barricades and, crouching behind them,
held the nozzles of the hose in position until the fire came down
into the streets and seized upon their shelter. They risked their
lives on precarious projections and hung to the roof and window-
sashes, using one hand for self preservation and with the other
giving direction to the streams of water. Their clothing was
seared, hair singed, faces discolored, hands blistered, lungs cau-
terized by the heat; and yet they flinched not."

Sometime between 11 P.M. and midnight, Boston Mayor
William Gaston, now thoroughly alarmed, called for Damrell to
leave the fire lines and meet with officials. Many citizens and

officials were swarming Boston City Hall to complain that the firefighting was chaotic and disorganized and Damrell was not in command of the situation. Indeed, a rumor was already swirling through the streets that the chief had cracked up and was sent to the insane asylum. As flames licked the Federal Street building in which they gathered, "Mr. Chief, what are your plans and purposes for the staying of this fire?" Gaston demanded of Damrell. Damrell, Gaston said, seemed calm and collected, although the fire outside was "a scene of confusion." Damrell described his deployment and tactics and said he was prepared to use gunpowder, even though he considered it a dangerous last resort. Gaston gave his approval, and Damrell and his engineers turned to the fire lines.

Damrell's chief critic was the city's postmaster general, William L. Burt. With his imposing beard and military posture, Burt was heralded in newspaper accounts immediately after the fire as a man of action. He went to the fire at the striking of the first alarm, worried about the threat to the nearly completed post office, and watched the devastation spread, growing angrier by the second at what he considered the lack of organization. He saw the situation as a "battle with fire." "The fire was then getting so large that it required a systematic defense, a systematic resistance, and no such defense appeared to be made," he said later. Burt was convinced that only strategic gunpowder explosions could halt the fire. While mail had been removed from the post office building on Devonshire and Milk Streets to the safety of Faneuil Hall, Burt said he'd be damned if he'd see his new building, supposedly built to be fireproof, consumed without a fight.

Thus at about 2 A.M. he and other citizens called for another meeting, this time at City Hall, to discuss gunpowder. Damrell left the fire lines with great reluctance, believing he was called by the mayor. Tersely, the mayor asked Burt what he wanted. "The fire is destined to sweep the whole city, unless there is organized resistance," Burt declared. "We must blow up buildings, the firemen are falling back, they are losing confidence and energetic action is needed."

The mayor demurred, saying that these decisions were not up to him but to Chief Damrell. "I am ready to receive any aid that might be offered in that direction," Damrell said wearily. Pointing to the buildings that blocked the view of Boston Harbor, Burt spoke sternly, "Mr. Mayor, before tomorrow morning if you look out of that window (if City Hall is saved, which I doubt, unless something is done) you will see the shipping in the harbor." Stung, the mayor retorted, "Would you take the responsibility of blowing up buildings?" It was the opening Burt was waiting for. Of course he would. Damrell was reluctant. He strongly opposed the use of gunpowder. He knew, however, that he was running out of options and that he needed assistance. Moreover, he was anxious to get back to the fire lines. He wrote out notes intended to give citizens permission "to assist in removing items from buildings and to blow them up." With the air of a general at a parade, Burt told Gaston and Damrell, "I shall go down to Winthrop Square and blow up the entire block between Federal and Devonshire."

He proceeded to do exactly that. Other officials also began setting off explosions, using gunpowder brought in from private stores and military posts. Among them was General Henry W. Benham, a Civil War veteran in charge of forts on Boston's harbor islands, who had come to city hall to volunteer his services. In fact, Benham had told Damrell not to return to the fire lines but to set up headquarters in a safe area. "That will do, General, for the field but will not do in this case," Damrell said before he went back into the streets.

Whether Damrell's notes gave Burt and Benham the authority to blow up every building they pleased was a point of much dispute after the fire. Gaston, not unlike many modern politicians, developed memory loss about the meeting, saying only that he had cautioned against endangering human life. "General Burt," he added dryly, "talked a great deal." Burt later proudly declared he had no idea of what the law said concerning gunpowder, indeed, he made it a point not to. "On such occasions as that I think no man should stop to study the law."

In addition to the threat that it might strengthen rather than stop the fire, gunpowder posed another danger. The natural gas system that provided lights to homes and businesses could be shut off only building by building; the gas pipes lacked check valves that could halt the flow of gas street by street. A number of firefighters risked their lives to shut off the gas in buildings. Now the explosions were igniting huge gas fireballs. With bizarre relish, Burt described the sight: "I saw in the most intense part of the fire, huge bodies of gas; you might say 25 feet in diameter—dark opaque masses, combined with the gases from the pile of burning merchandise—rise 200 feet in the air and explode, shooting out large lines of flame fifty to sixty feet in every direction, with an explosion that was marked as the explosion of a bomb."

Burt afterward insisted gunpowder was the salvation of the city. "The roar of the explosions was the first sign of hope and sent a thrill of joy through the entire city." Indeed, Sarah Putnam wrote, "At last, about 3 o'clock, some buildings were blown up and everyone rejoiced to hear the explosion for it seemed as if something would check the flames." While those outside the fire zone were cheering, inside firefighters feared for their lives. Gunpowder charges were not set in any systematic fashion; sometimes they failed to go off at all or often just blew out a building's windows without taking it down. An ex-alderman who helped set a charge was not even sure the explosion took down the building because "I was running for my own preservation."

Damrell blew up two buildings near the corner of Batterymarch and Milk Streets. With the help of Captain Jacobs, he hauled ten 25-pound kegs of powder upstairs into one of the buildings. Brushing sparks and cinders off the kegs, he lit the fuse, saying to Jacobs, "If we go up, we will go together, we will make a clean thing of this." Damrell and Jacobs barely made it to safety before the explosion brought down the building. The fire, however, simply swept right over that building to the next one.

By early Sunday Damrell's engineers were begging him to halt the gunpowder explosions. Damrell decided the experiment had

Washington Street before. . .

gone far enough; he sent a captain out to order all explosions halted. He also sent word to Alderman William Woolley, head of the Committee on the Fire Department, who took the order to mean: "If anybody undertakes to blow up any more buildings, have them arrested, and if you can't arrest them, kill them."

Woolley did not hesitate to follow this order. He opposed gunpowder, and he personally prevented dozens of explosions through the night. In a scenario of dark absurdity, he almost came to blows with General Benham, who was seeking to blow up buildings between Kilby and Congress Streets. Benham insisted he was one of the men granted permission to oversee the

. . . and after the 1872 fire

gunpowder explosions. Woolley, however, insisted that no explosions were to be allowed.

"Who gives you such authority?" Benham sputtered.

"I am the chairman of the Committee on the Fire Department and these explosions are demoralizing the firemen," Woolley shouted.

"Do you know who I am?" Benham demanded.

"I do not."

"I am General Benham."

"I don't care if you are General Damnation. That powder is not going into the building." Firefighters left their lines to watch as the two men squared off. Woolley eventually gave in, but the arrival of Captain Joseph Dunbar with orders from Damrell halted any more explosions.

By 4 A.M. fire had gutted the blocks between Franklin and Milk Streets and dealt a terrible blow to Boston's publishing industry. Consumed were buildings for the *Transcript*, the *American Union*, the *Saturday Evening Gazette*, and *The Pilot*, Boston's Catholic newspaper. Instead of attending services on Sunday morning, Bostonians flocked to the burning zone, leaving churches deserted for miles around. In the hours before dawn, Boston's Episcopal bishop, the Reverend Phillips Brooks, watched the destruction of Trinity Church at the corner of Summer and

The battle to save the Old South Meeting House

Hawley Streets. Brooks, author of the carol "O Little Town of Bethlehem," had thought the church was safe through the night, but toward the morning he and the staff started carrying out books, robes, and religious objects. "She burnt majestically," Brooks wrote. "I did not know how much I liked the great gloomy old thing till I saw her windows bursting and the flames running along the old high pews."

The fire devastated Boston's leather goods and wholesale cloth industry. Eight million pounds of wool went up in smoke. An estimated $10 million in consigned goods, 550 separate estates, and nearly 1,000 firms were burned. Precious paintings and rare books became kindling. The flames took paintings by John Singleton Copley and Anthony van Dyck, 10,000 volumes from the Reverend John Singleton Copley Greene's library, and ancient weaponry collected by Colonel T. B. Lawrence, including seven complete suits of armor and "dreadful instruments of torture hideously suggestive." Charles Carleton Coffin, in his eyewitness account of the blaze, found no superlative too great to describe the fire "giant": "The fire leaps from every window, rolls down the spacious stairways in billows of flame, entwining the granite

The Worcester Fire Department arriving to help fight the Great Fire

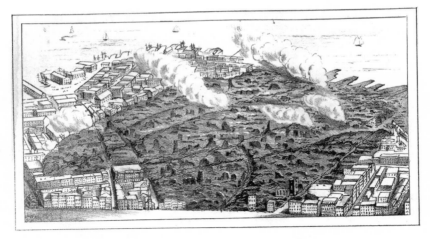

Boston in ruins after the fire

pillars, hanging in red folds, beneath the cornice. The richest fabrics of European looms—silks, velvets, satin, lace, and ribbons—one hundred thousand dollars' worth a minute vanish in smoke and flame."

Worst of all was the threat to the Old South Meeting House, one of the last remaining colonial structures. Here Sam Adams planned the Boston Tea Party; here Bostonians had gathered to debate issues of the day; Lyman Beecher had even given one of his notorious anti-Catholic sermons here. Now fire was at its door. Firefighters poured streams of water on its walls, and daring souls climbed onto the roof to sweep away sparks. Burt dismissed three demands that the building be blown up. The battle raged through the night, and when the steeple clock struck 6 A.M., a bystander exclaimed, "Dear old church, I'm afraid we shall never hear that bell again." In the nick of time, an Amoskeag steam engine, Kearsarge No. 3, from Portsmouth, New Hampshire, arrived. The steamer had been loaded on a flatbed train with the Portsmouth company and taken into Boston. The fresh New Hampshire firefighters fired up the engine and ran to the fire lines. At the Old South, they hooked to hydrant and trained the hoses toward the roof of the church. Their efforts turned the tide. The fire was stopped at Washington Street and the Old South survived.

The last line of defense became State Street, just outside the city's old post office. More than forty steamers massed along State Street and poured torrents of water on the buildings. The heat was fierce, but the line held. By 1 P.M. the fire was deemed under control. By 2 P.M. out-of-town steamers were being sent home. But the "fire fiend" was not yet sated. Against all reason, the city's gas mains continued to flow. About midnight Sunday, a gas explosion leveled a building on the corner of Summer and Washington Streets, sending the iron manhole covers sailing through the air like autumn leaves. Flames quickly engulfed a building housing Shreve, Crump & Low. A panicked woman jumped from a window; her mother was later found dead in the ruins. Finally, all the city's gas mains were shut off, and Boston was plunged into darkness for two days. "The gas has been shut off, so we have only candles & no street lamps, and it is altogether so melancholy that I should like to have a good cry," Mrs. Grace Revere of Boston wrote a friend in Europe. Firefighters continued to play streams on

Another contemporary engraving in Harper's Weekly, *of a terrified woman during the fire*

this block until Monday morning. At last, the fire was officially over. A weary Damrell, his lungs singed by the heat, his energy taxed to the limit, made his way home.

RISING FROM THE ASHES

As Bostonians looked over their devastated city Monday morning, many compared the sight to the ruins of Pompeii or Carthage. "It will be remembered as the great fire, the greatest that America has seen, with the exception of that which laid waste Chicago," Charles Carleton Coffin boldly asserted. Block after block was a charred, smoky pile of rubble. Scorched columns of once magnificent granite structures jutted from the earth like bones of felled mammoths. The new post office survived. "It was like a fortification," Burt proudly declared. Old South was saved, but 776 buildings were gone and more than sixty acres were burned. The total damage was estimated at $75 to $76 million—which would be well over $500 million today. Thousands were left homeless, 20,000 jobless. About 20 of 33 insurance companies went bankrupt attempting to cover losses. "The desolation is bewildering," the Reverend Brooks wrote.

Nine firefighters—from Boston, Cambridge, Malden, and Worcester—and one former firefighter were dead; two more firemen soon succumbed to wounds. The widowed mother of Walter S. Twombly of Hose 2 Company of Malden searched "through the ghastly piles of smoking debris for some sign by which she might find the body of her son," Conwell reported. Sixteen other people, including two children, were reported killed or missing, but an exact toll of the dead and wounded was never officially determined. Damrell never did learn the fate of the parents of the lad who begged him for help.

The state governor called out members of the Claflin Guards and other military companies to assist the police in keeping order. Rumors about looting and invasions of criminals were repeated and embellished: "I met Mrs. Edmond Codman, who said that a man had been caught in the attempt to set fire to a house and that he had been shot," Sarah Putnam recorded in her diary. "We have just heard that two men were caught [looting] in the night and hung on [a] lamppost which seems too small a punishment for such scamps," Grace Revere wrote. While these rumors were merely rumors, many Bostonians fearfully prepared for the worst. Police Chief E. H. Savage readied his department upon hearing reports that "toughs" from Springfield were on a train bound for some Boston looting. They never arrived. Savage said grimly: "If they had, we should have had some fun with them." Murdoch's observer had a more a jocular view: "I got out my father's old revolver which I think had not been loaded ever since 1863, and if I had been called upon to fire it, I suppose it would have destroyed the house and everyone in it."

One of the many advertisements appearing after the fire, announcing that the business was "not burnt out"

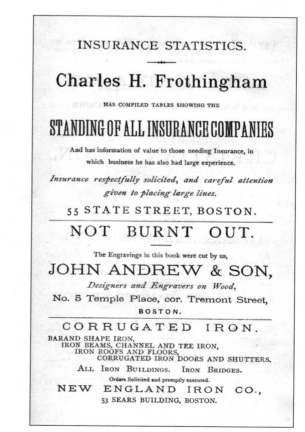

INSURANCE STATISTICS.

Charles H. Frothingham

HAS COMPILED TABLES SHOWING THE

STANDING OF ALL INSURANCE COMPANIES

And has information of value to those needing Insurance, in which business he has also had large experience.

Insurance respectfully solicited, and careful attention given to placing large lines.

55 STATE STREET, BOSTON.

NOT BURNT OUT.

The Engravings in this book were cut by us,

JOHN ANDREW & SON,

Designers and Engravers on Wood,

No. 5 Temple Place, cor. Tremont Street,

BOSTON.

CORRUGATED IRON.

BAR AND SHAPE IRON,
IRON BEAMS, CHANNEL AND TEE IRON,
IRON ROOFS AND FLOORS,
CORRUGATED IRON DOORS AND SHUTTERS.
ALL IRON BUILDINGS. IRON BRIDGES.

Orders Solicited and promptly executed.

NEW ENGLAND IRON CO.,

53 SEARS BUILDING, BOSTON.

Many Bostonians used the fire as an excuse for intoxication. Blocks of the city turned into scenes of macabre frivolity as citizens drowned their sorrows. This included firefighters, some of whom had been given nothing to eat or drink for a day except a tumbler full of whiskey. "The number of tipsy men seen in the neighborhood of the fire baffles computation," Conwell reported. To curtail the carousing, Savage ordered all drinking establishments closed. As the city faced a dry week, Murdoch's observer ran into a man who said grumpily, "What's the use of having such a h—l of a fire in Boston if a feller can't get a drink?" Added Murdoch, "I thought the gentleman's point of view very interesting. Some of the clergymen tell us that the fire was sent as a warning to us to mend our ways, but here is a man who holds other views and who asserts that the fire has utterly failed of its object."

Certainly the clergy wasted no time in drawing on the fire for sermons. "The sun went down last night smiling upon a great and prosperous city; when it rose this morning, it glared upon a roaring storm and flame; and tonight it sets on a wilderness of ashes," Henry Ward Beecher, son of Lyman Beecher, told his congregation. Unlike his Puritan forebears, Beecher did not evoke a wrathful God. On the contrary: "Instead of asking if God meant to humble Boston, let us look to the future and see what are lessons to be learned from such a conflagration as this." Other Bostonians regarded the devastation with dark humor: "Trinity Church, its tower standing, its walls partly fallen [is] more imposing as a ruin that it ever was in its best estate," Dr. Holmes wrote.

Soon merchants and the curious were venturing into the still smoking rubble to scavenge what they could. Safes were dragged from the smoldering ruins; some held only cinders, but more than a few still held precious documents and money. The Bank of North America recovered over a $1 million worth of securities and money from scorched but intact safes. Boston Common, the *Boston Journal* reported, remained "one vast depot of rescued goods." Newspapers feverishly rushed out extra editions, and artists churned out lurid drawings of burning buildings, heroic firemen, and the dramatic rescue of Old South. Out-of-town papers mourned the destruction in the nation's Revolutionary War capital, although a New Hampshire daily sniffed that the people in Boston generally were a "very self-satisfied lot." Photographers, eager to exploit the then-new technology, rushed to the scene. While no photos of the actual fire exist, hundreds of photos were snapped of the ruins from every possible angle. Demand was so high that one photographer wrote he had "hardly time to take a breath, such has been the call for photographs of the burnt section." Stereographs were then popular, and Boston's ruined landscape became parlor entertainment, planting the seeds, perhaps, for the first Hollywood disaster movies. Chandler and Company rushed a booklet on the fire into print in ten days. Its final pages included an ad for Morris and Ireland safes, with gushing testimonials from the grateful owners: "A safe that fell from 2nd floor to cellar amongst hot bricks was opened Monday, everything in it was all safe even of the loose papers not one being so much as scorched." Even this calamity, it seemed, had commercial potential.

Armageddon could not quench the ingenuity of the Yankee merchant. Signs soon appeared on smoldering ruins: "Removed During Repairs," "Gone Up," "A Burning Shame," and "Closed During the Heated Term." Boston's *Saturday Evening Express* reported that the Tuesday after the fire, the city's five daily papers contained advertising from 310 burned-out firms about their new quarters. Unlike the Chicago fire, which destroyed residential areas, Boston's fire was concentrated in its commercial district. "But for the girls thrown out of employment, the loss is mostly confined to those who are able to stand it," Joslin wrote his brother. "Folks do not appear much dejected over it, owing, I suppose to knowing that all their neighbors are in the same predicament." Patrick Donahoe, editor of the devastated *Pilot*, sighed and said, "Well, it's gone, but we'll resume business at the old stand tomorrow." The *Transcript* managed to put out an issue on Monday by printing at the *Boston Globe*'s plant. Thanks to Burt, the mail was delivered with only minor interruptions. "Two merchants who were one week ago worth $30,000 each were yesterday 'waiting upon table' in a North End restaurant," the *Saturday Evening Express* sarcastically noted.

Wealthy Brahmin life hardly missed a beat. When Sarah Putnam's cook went out to look for relatives caught in the fire, she returned the next morning "drunk and dirty and tearful. Of course, I had to send her away though I felt quite badly as she had been so neat and pleasant." As families looked for homes and store owners tried to calculate losses, "I have been hunting up a cook," Putnam wrote. Her brother, J. Pickering Putnam, a prominent architect, wrote a condescending letter to the editor of the *Boston Daily Advertiser* in defense of the now universally despised mansard roofs: "It is not the form of the roof which is are fault, but the material and character of construction."

"True," one reporter noted, "there were numerous instances where men who saw the accumulation of a lifetime vanish in smoke in an hour, cried like children and gave way to despair; but the great masses who thronged everywhere were never for a moment panic-stricken or palsied with fear."

The mayor of Chicago, which had received help from Bostonians after its great fire a year earlier, sent a telegram saying simply, "In what way can we help you most?" A relief fund for the families of firemen killed and for the thousands left homeless or unemployed (or both) was quickly set up; $80,000 was raised for firefighters' relief. Abolitionist writer Harriet Beecher Stowe contributed $100, writing, "No soldier that died for our common country deserved greater honor."

But someone had to be held responsible, and Damrell found himself the focus of wrath and derision. Newspapers lauded Burt's gunpowder explosions and sneered at Damrell's cautiousness. "It would have been well had the city been blessed with a few more of the same sort of courageous determined men (like Burt) in the midst of this terrible exigency—although , unquestionably, every one in authority did 'his level best' in this trying hour," Chandler's reporter opined. The *Saturday Evening Express* sniped, "There don't seem to be any sick horses around excepting those belonging to the fire department. If they had been

*Helmet worn by
Chief Damrell,
now on display at
the Dover Fire
Department*

permitted to inhale smoke last Saturday night the fire would have been stopped." Firefighters, however, rushed to praise Damrell: "The ability, coolness, indomitable courage and perseverance displayed by chief engineer Damrell in his efforts to arrest the progress of the fire merits and receives our earnest and unqualified approbation," declared a resolution passed on November 22 by a group of the state's chief engineers.

This being Boston, a commission was appointed to investigate the cause and spread of the fire. From late November 1872 to early January 1873 five commissioners listened as 230 witnesses—ranging from Damrell to Burt to Woolley to Jacobs to Eliot to police, firefighters, and passersby—recounted their versions of events. Commissioners grilled Damrell unmercifully, but the chief welcomed the chance to repeat his warnings about the area's water mains. This was, he said emphatically, a fire like no other. "I have been to oakum fires and oil fires and lead fires and fires of flour mills, and everything of the kind,—cotton fires, but I never saw anything like this, and I don't think there is a man in the Fire Department who ever did." As for gunpowder, "Never again would I use it."

Burt recounted his deeds with relish, taking digs at Damrell whenever he could. John Bird could not have been happier to read from his (recently reprinted) article on mansard roofs; the commissioners had to hustle him along when he began to describe his own antifire devices. Eliot, apparently believing that a Harvard president was also a fire expert, insisted firefighters worked nonchalantly and "without desperation" and described how he tried to tell Damrell to station firefighters on roofs to better fight the fire, although, Eliot acknowledged, several firemen pointed out the extreme danger of getting on those roofs.

"This danger had been foreseen and our calamity had been foretold both here and abroad," the commissioners lamented in their final 662-page report, which included all witnesses' word-for-word testimony. As to the cause of the blaze, "no answer could be made." The commissioners did find other culprits: the mansard roofs, the sick horses, the water mains—and the chief engineer. Rather than being fought with an overall coordinated effort, "the fire was attacked piecemeal as chance occurred." While "the Chief Engineer deserves all praise for his courage," Damrell unfortunately "tried to unite the services of a private with those of a commander in chief." The commission adroitly sidestepped the issue of gunpowder's effectiveness, saying only that explosives should never again be used the way they were in the great fire. The group recommended that the fire department stock up on a new, better explosive called "dynamite."

The commission did suggest that the city install a better water distribution system in the burned district. But only after a serious fire in February of 1873, which killed three firefighters, and another on Memorial Day, May 30, 1873, which destroyed 105 businesses, did the tightfisted Yankee establishment launch a major overhaul of the water mains and shift the fire department from a political to a professional sphere. A three-person board of fire commissioners was established to appoint and oversee a chief engineer, who would, in turn, appoint assistant chiefs. The new fire commissioners also took over fire-fighting functions previously delegated to various city committees. Damrell's vision of a fully professional fire department was becoming a reality—without him. He was replaced as chief engineer in 1874.

In the meantime, builders and merchants got to work. Rubble was hauled away, and the narrow streets in the burned area were widened and straightened, including Devonshire, Franklin, and Milk Streets. Land was cleared around the new post office for the open space of Post Office Square. In two years virtually no trace of the fire remained. Today only a couple of historical plaques and the infamous Box 52—still in the same location—provide any evidence of the conflagration. The wharves are gone, landfill has now changed the harbor's edge, and skyscrapers tower over the still narrow streets, which were mostly rebuilt along their original configurations. The Old South Meeting House still welcomes tourists, but Burt's pride and joy, his "new" post office, is gone, replaced by the towering John W. McCormack Post Office and Court House, built in 1931. Designed by H. H. Richardson, Trinity Church was rebuilt in the then "new" area of Back Bay and is heralded as an architectural masterpiece. Department stores now fill the blocks where Trinity Church once welcomed parishioners and *The Pilot* and the *Transcript* once churned out editions.

By the fiftieth anniversary of the fire, the number of Boston firefighters had grown to 1,400 men—all permanent. In 1881 a new "keyless" system of sounding fire alarms was installed in the business district, replacing the need to find a policeman or responsible person to trigger them. Another innovation came that year: all alarm boxes were painted red. New building codes "brought into use many devices like automatic sprinklers and water curtains, automatic doors, wire glass and automatic fire alarms to prevent and retard the rapid spread of fire," Charles Damrell, the chief's son, wrote in 1922. In 1872 it took nearly two hours to assemble twenty-one steamers; "the present chief could mass almost fifty pumping engines on Boston Common in less than half an hour," the younger Damrell noted.

Yet the Great Fire of 1872 continued to haunt Boston's fire chiefs, a lingering ghost of doubt in the back of their minds that whispered, "Could it happen again?" On the 100th anniversary of the fire in 1972, District Fire Chief John Vahey, troubled by that question, completed a thorough study of the disaster he had dubbed the "Epizootic Fire." Damrell, he decided, did everything he could possibly have done. A confluence of events—the initial delays, the low water pressure, the unchecked gas flow—conspired to make what could have been an easily fought blaze into an unstoppable inferno. Only continued vigilance would ensure that Boston would never again have a fire of such magnitude, he concluded.

John Damrell, though castigated, was not disgraced. In 1873 he helped found the National Association of Fire Engineers (now the International Association of Fire Chiefs) in a Baltimore convention and was elected its first president. In 1877 he returned to his first love—preventing fires before they start—when he was appointed to the newly created position of building commissioner, where he served for the next twenty-five years. His son Charles became clerk of the building department and helped rewrite and update its codes.

John Damrell lived to see the dawn of the twentieth century; he retired in 1903 and moved permanently to his country home in Dover, Massachusetts. After his death on November 3, 1905, his obituary read, "He has been conceded to be a master of the extinguishment of fires and an expert on advanced ideas connected with that important service." Two of his helmets remain on display at the Dover Fire Department; the leather is worn, the white paint nearly faded away, but the crown still proudly bears the name "Damrell" below a still-colorful rendition of a steam engine.*

Unlike the Chicago fire, which grew in legend, Boston's fire faded from memory. Few Bostonians, rushing to banks and offices in the financial district or shopping in Downtown Crossing, realize the devastation wrought there 160 years ago. Boston's Great Fire of 1872 was costly, but the loss of life was minimal, and nothing hits a city harder than the loss of its citizens. Boston would learn that lesson seventy years later on another Saturday night in November.

---

*The obituary also makes cryptic references to a three-year moratorium on contracts in Damrell's private business by "reason of attachments on account of his connections with the explosions of buildings with powder at the great Boston fire in 1872." Damrell was, perhaps, still held accountable by some for the explosions he had tried so hard to prevent.

6

TWICE BURNED IN CHELSEA

The 1908 and 1973 Conflagrations

Firefighters don't lightly toss out the word *conflagration*. An average citizen might think the term just refers to a "big fire." To an experienced firefighter, a *conflagration* means a fire that has jumped a street, either when buildings are closely situated or when radiating heat or flying embers ignite nearby structures. A conflagration is thus a very serious fire in danger of raging out of control. Boston's fearsome 1872 fire was a conflagration. So were many of the colonial blazes that wiped out whole blocks of homes. But as firefighting equipment improved, alarm systems grew more efficient, and firefighters were better trained and equipped, the word *conflagration* seemed to become a quaint expression of old, a term that hearkened back to the days of hand pumpers and steam engines.

Which may be why the dispatcher for the Intercity Mutual Aid System was initially confused by a radio call from the fire chief of Chelsea in the late afternoon of October 14, 1973. "A confla-what?" he radioed back.

"Notify Newton Control that Chelsea has a full-fledged *conflagration*," Chief Herbert C. Fothergill barked into his radio. Meaning: Entire blocks of Chelsea were now roaring furnaces, and if the fire departments of other cities and towns didn't pitch in soon, the city would go up in smoke.

Perhaps the dispatcher's confusion was understandable—Fothergill's call was the first time in New England that a conflagration had been called in by radio. Likewise, Fothergill's impatience was understandable. Every Chelsea firefighter had been reared on tales of Chelsea's 1908 conflagration, a Palm Sunday inferno that wiped out a third of the city, killed nearly twenty people, and destroyed 3,500 buildings. Some of the firefighters could even recount stories of that fire told by their fathers or grandfathers.

Now, after sixty-five years, a conflagration had started in almost the exact same spot, under conditions eerily similar. Just as his predecessor, Chelsea Fire Chief Henry A. Spencer, had quickly realized in 1908, Fothergill saw that a fire had jumped a street and was turning into a bad blaze—a very bad blaze. So the call went out to fire departments around the region with the message: Chelsea is burning. Again.

Palm Sunday, April 12, 1908, dawned dry and dusty in Chelsea, then a city of about 38,000, bordered by the Mystic and Chelsea Rivers, just across from East Boston and Charlestown. By the early 1900s Chelsea suffered from a reputation as a factory-filled poor community, although its proud citizens railed against the expression "as dead as Chelsea," which was then common. Chelsea, they noted, was as old a community as Boston, with its own distinctive history. The area was first settled in 1625 by Samuel Maverick and became part of Boston; during the 1776 siege of Boston, when revolutionary troops were stationed there, Gen. George Washington paid a visit. On January 10, 1779, the town of Chelsea was split off from Boston; the town's boundaries were later redrawn in 1841 and again in 1846. When the population swelled to 12,000 in 1857, the town was granted a city charter. Readily accessible by ferry from Boston, Chelsea was then considered an aristocratic suburb of Boston, with gracious tree-shaded streets, fine houses, and a bustling commercial area. Bostonians might sniff about being "as dead as Chelsea," but the city boasted many fine houses, schools, churches, synagogues, grocery stores, and tailor shops. In short, Chelsea was a diverse, bustling community—with an Achilles heel.

To this day, Chelsea fire officials believe Boston's 1872 fire contributed indirectly to the 1908 Chelsea blaze. To prevent another such conflagration, potentially hazardous businesses, such as paper- and rag-reprocessing businesses, were sent packing from Boston, across the Mystic River, into Chelsea. The so-called rag district soon spread from the waterfront to the railroad tracks that cut through the middle of the city to Arlington and Williams Streets. The area filled up with two- and three-story wooden buildings, housing about 200 rag and junk shops as well as a variety of small factories. Also, the waterfront became a mecca for bulk oil storage terminals, as the demand for heating fuel, and later motor fuel, grew. In the late nineteenth century, immigrants from Eastern Europe began arriving, including many Jewish families fleeing the harsh rule of Russian czars. Many found work and housing in the rag district. Tenements became packed with the extended families of the newcomers, many of whom launched small businesses of their own. On some streets nearly every shed, stable, and yard contained rags, according to a report by the Underwriters' Bureau of New England, which concluded, "The consequence was that the district was a conflagration breeder of the worst kind." City officials were aware that the district posed a fire risk, yet they did nothing about it. At that time the fire department had seven fire companies, with twenty-one permanent and fifty-seven call men.

The winds on that Palm Sunday were averaging twenty-three miles per hour, with gusts up to thirty-nine miles per hour. At about 10:44 A.M. Box 215 was struck; firefighters instantly knew a fire had broken out in the rag district. Chief Spencer and one of his companies raced to the scene at Boston Blacking Company, near the intersection of Carter and Summer Streets. They found a fire rapidly consuming a rag dump near the company, and within minutes the firefighters handily had it under control. Then: "Chief, look up there!" someone cried. Spencer turned and saw flames licking the roof of the Rosin Storage building, possibly spread by flying embers from the original fire. He called for a second alarm. Just fifteen minutes later, and several blocks away, fire was reported in the rear of a rag house on Maple Street. Witnesses said the entire second story seemed to burst into flames.

In a few minutes the firefighters realized they were battling no ordinary blaze. Fire was reported at the synagogue on the corner of Elm Street and Everett Avenue, about a quarter-mile away, and at the corner of Walnut and Fourth Streets. The shingled roofs of the tenements easily caught fire or made effective launching pads for embers to be blown elsewhere by the wind. Showers of sparks were landing on rag piles and paper stacks throughout the neighborhood. Spencer soon realized that the fire was outrunning his men, and he sent a general call for help to the surrounding cities. All through the afternoon, horse-drawn steamers galloped in from Boston, Everett, Revere, Lynn, Winthrop, Cambridge, Malden, Medford, and Melrose. In the meantime, however, the fire was moving rapidly through the city and turning into a firestorm, in effect, a hurricane of fire that could generate its own winds of up to 100 miles per hour.

Desperately, Spencer and his men tried to find a wide-open area from which to set up apparatus and hose lines and hold back the advancing fire. But again and again they were forced back, as the fire pressed forward, stoked by the wind and the readily available fuel: rags, tinder-dry homes, even furniture and bedding abandoned in streets. One of the last companies to leave their post on Bellingham Street was Lynn's Engine Company No. 1. With their faces burned and blackened and their hair singed, they would not be beaten back, not as long as they could keep their pump streaming water. But the water was turning to steam in midair, and soon it was beyond all human ability to withstand the terrific heat and suffocating smoke. The Lynn men had pushed their luck; the flames had now penned them in. Fleeing for their lives, they abandoned their beloved steamer and escaped through the billows of flame. According to legend, the company's captain sat on a curbstone a safe distance away, tears running down his cheeks. "Listen, boys. Hear our steamer? She's still pumping. Hasn't she got a nice beat?"

Frantic residents were gathering up their belongings and attempting to flee. Eyewitness Walter Merriam Pratt, from one of the city's prominent families, wandered the streets, mesmerized by the unfolding drama. "It seemed as if everyone tried first to save a mattress which would become ignited before it was carried a block, and add to the column of the flame," he later wrote in the sole book on the fire, *The Burning of Chelsea*. "In some cases men and women fought as to what they would save, while their houses burned."

Union Park was soon piled with people and their goods, but the choking smoke quickly drove the people out, and the fire took their possessions. The pattern was repeated throughout the afternoon as people reached a place of supposed safety, only to have to move again. "No one seemed to realize how fast the fire was moving except those who fought it," Pratt observed. As at any fire today, onlookers, who were even then dubbed "sparks," came to gawk. One of them later told a newspaper:

> *There was confusion everywhere. People were beginning to move furniture and household goods out into the streets. They were crying, screaming, looking for lost children, relatives, and pets. Nobody seemed to know what to do or how to do it. We stopped at numerous houses to warn residents to seek safety. As many of those people did not speak our language, we had difficulty with them. In one such situation, with the roof afire, we had to forcibly drag an elderly couple out of the house. A neighbor took them in charge.*

The parade of people and possessions was tragic and bizarrely humorous. Hundreds carried out cages of canary birds; one poor woman carried a canary cage with a cat inside, never realizing the cat had eaten the bird. Another woman carried a great marble clock under one arm and a dog under the other. Two men trying to save an upright piano gave up when the cloth on the back started to burn. One opened the lid and played a few chords of "There'll be a hot time in the old town tonight" before abandoning the instrument to the inferno.

In the People's AME Church on Fourth Street, Pastor Charles P. Watson was leading the service; parishioners had no idea of the inferno bearing down on them. Fortunately, someone ran inside to warn the reverend that the church was in the path of the fire and could not possibly survive. Calmly, the Reverend Watson told his flock that the fire was spreading, led them in a hymn, and then proceeded to evacuate.

To fleeing residents it seemed the apocalypse had come. Said one newspaper account:

> *The flames seemed determined to destroy the city and acted as though possessed of an evil spirit. It was impossible then to stop the fire. The monster seemed bound to lay waste the city. Foot by foot, but none the less steadily, the firemen were driven back. They made a heroic stand but all in vain.*

Another reporter wrote:

> *The fire brooked no interferences. It darted down Spruce Street and crossed the railroad tracks, swallowed up several buildings that stood in its path, refused to destroy several lumber yards that seemed to stand directly in its path and then seemed to rush up the side streets with such speed that only the waters of the Mystic, more than a mile away, halted the progress of the conflagration.*

Rich and poor were on equal terms in the maelstrom. Mrs. J. B. Fenwick, her niece, and her maid fled their fine house on Chestnut Street but were trapped by falling buildings and took refuge on a porch, where they were overcome by smoke. Their bodies were found four days later. The magnificent brick and stone mansion of ex-mayor Thomas Strahan, at the top of Bellingham Street, was caught in a seething mass of flames that moved like a tidal wave up the hill. Horses galloped through the streets, and the hurricane-force winds, generated by the fire, knocked men off their feet and sent huge pieces of furniture bounding end over end. In City Hall, the city treasurer removed all the city funds and locked books in the safes, and the city clerk stayed in the building so long that he had to escape through a second-story window before the flames roared in.

"The wails of hundreds of parents vainly searching for their children added to the excitement," Pratt reported. "One mother fell in a dead faint when her two-year-old child, whom she had given up as lost, was brought to her." Pratt had an eye for the small details of the fire: the man who rushed into his house to save his cat, only to have it run back inside the burning house; the fireman who entered a burning drugstore and took his reflection in a long mirror for another fireman and walked into the glass. "All at once out of Hawthorne Street shot an engine as if coming out of a cannon. The driver was almost doubled up and the horses were going at a two-twenty clip; where they came from or how they ever got out of that furnace alive is a mystery." Another observer noticed that, once safe, the man jumped off his rig and quickly went around his horses, checking them for injury. He stroked them and hugged them around the neck, as much as to say, "Boys, you saved our lives."

Eli C. Bliss lived near the railroad tracks, and as he anxiously watched the fire drawing close to his home, a passing freight car chugged to a halt. "We have two empty box cars," a railman cried. Working like demons, Bliss and friends carried everything of value from his house—including a grand piano—and hoisted it into the freight car. The train pulled out just as the house was engulfed in flames. Bliss was able to retrieve all his belongings later.

The railroad tracks that cut though the town proved a blessing in another way; they created a fire break, allowing firefighters to save the northern end of the city. At the southern end, the fire burned steadily toward the oil tankers on the waterfront. As the tanks caught fire and exploded, sending plumes of flame hundreds

Chelsea Fire Chief Spencer with his eyes wrapped as in the 1908 conflagration

of feet in the air, flames ignited the docks and reached three barges loaded with oil. Soon Chelsea Creek was covered with burning oil, carried by the incoming tide to wharves of the Standard Oil Company on the East Boston side. A fireboat was hemmed in but managed to direct its streams into creating a safe passage for itself. Throughout the afternoon East Boston firefighters dashed from one hot spot to another, as burning embers set fire to roofs. With six engines and lines laid through the district, "it was only by an efficient and determined fight that East Boston was saved," Pratt declared.

Their necks and ears seared by the ferocious heat, firefighters in Chelsea struggled throughout the day to set hose lines and

Bellingham Methodist Church and Old High School, Chelsea, 1908

hold back the flames. "Give me hell in preference to this," one groaned. None, however, deserted their posts until driven out by heat and flames. "A great many Chelsea firemen lost their homes but even when it became apparent that they were to be destroyed, they didn't waver but stuck to their duty," Pratt reported. When an assistant chief was told his Essex Street house was ablaze, he calmly went home, secured a few clothes and papers, and—with his two sons—left the place prey to the flames and returned to duty. Chief Spencer, his eyes singed by ashes and embers, wrapped a bandage around his face and carried on.

Aided by the dying wind, the waterfront, and the railroad tracks, firefighters were finally able to make a stand. By 5 P.M. the fire had traced an outline of destruction and was "simply burning on the outskirts and finishing the buildings in the interior," as the chief's report put it. It wasn't until well into the night, however, that the final flame was doused.

"As the sky grew light and the morning mist cleared away, it displayed a vast expanse of smoking ruins," Pratt wrote. "The ruin is so complete, so disastrous, that figures can hardly give any meaning. Never before in New England have so many people been stricken so completely in one day with such appalling disaster," the

Chelsea in ruins after the 1908 fire

Boston Advertiser declared. All day, people streamed out of Chelsea into Boston, resembling a line of refugees fleeing a war. As the ashes cooled, stunned citizens and officials totaled their losses: The fire had cut a swath a mile and a half long and three-quarters of a mile wide—about 270 acres. Eighteen bodies were recovered, but more may have been consumed in the inferno. More than 300 injured were treated at local hospitals. Three fire companies, two from Boston and one from Lynn, lost engines when firefighters had to flee for their lives. According to Pratt's tally, the destruction included more than 700 businesses, 13 churches, 8 schools, 23 oil tanks. 3 banks, the public library, and more than 3,000 shade trees. Two fire stations were gone. So was City Hall. More than 17,000 people were left homeless. Total losses were estimated at about $20 million.

The fire brought out the best and the worst in people. The night of the fire, even as martial law was declared, hundreds of tents and blankets were brought in by a special train. A relief fund was established, and Massachusetts citizens ponied up more than $350,000. President Theodore Roosevelt telegraphed Chelsea Mayor John E. Beck: "In company with all our people, I am inexpressibly shocked

Powder Horn Hill

Ruins of
Williams School

Walnut St.

Ruins of the
Universalist Church

CHELSEA FIRE

Portion of a panorama of Chelsea, created not long after the fire

at the tragedy that has befallen Chelsea." Not even animals were forgotten: the Animal Rescue League sent people looking for homeless dogs and cats. The injured were at once chloroformed, the rest held for owners to claim.

The cause of the fire was never determined; deep suspicions that it had been deliberately set were never proven. Some believed a cigar carelessly thrown into a rag pile might have been the culprit, but no cause was ever officially pinpointed. Nonetheless, many, including Pratt, blamed the "Hebrew" rag shop owners. "The residents of Chelsea are determined to drive out the Hebrew junk dealers, and the insurance companies are helping by canceling all policies on rag shops. The people of Chelsea have tolerated these undesirable citizens as long as they propose to; fire after fire of incendiary origin has taken place until there is no alternative—they have got to go," Pratt wrote defiantly in the last pages of his book on the fire. Even the Underwriters' Bureau of New England, in its report on the fire, complained that "influential citizens of means" had fled the area, leaving "the foreign element of low moral stamina and small earning capacity." The kind of prejudice that once had been aimed at the Irish and a Charlestown convent now found a new target.

Slowly, the city rebuilt. The fine homes on the long, tree-lined streets were gone forever, and the waterfront was given up entirely to commercial use as oil tanks were rebuilt and expanded. Much of the rag district remained a zone for junk shops and waste recycling. Wood-frame tenements were covered with tin and converted into reprocessing shops. In a cycle seen in other urban areas as well, the Eastern European immigrants prospered, drifted north to more affluent communities, and were replaced by immigrants from Puerto Rico and Latin American countries. By 1973 the population had decreased to 30,000 residents, and the city had become one of the country's major salvage and reclamation centers.

Tales of the 1908 fire continued to be told by longtime residents. Fire Chief Herbert Fothergill tells how his father, Herbert Senior, witnessed the 1908 fire as a kid. When the lad disappeared as the city was burning, his mother was nearly beside herself with worry. He finally turned up, hours later, with a chicken under each arm. His mother promised him the worst punishment he had ever had, but the lad was unabashed. "I could have had a cow," he declared.

Firefighting ran in the blood of the Fothergills. Fothergill Senior became a fire captain. His son Herbert joined the Chelsea department in 1946, and he was designated chief in 1966—a tough era for any city official. As city budgets grew tighter in the 1960s and 1970s, efforts to avoid layoffs and obtain better fire equipment became political battles. Fothergill badgered officials about the lack of water in the city, particularly in the area still considered a "rag-picking" zone, since it was filled with combustible material—now tires and chemicals in addition to newsprint and cloth. The decades had brought innovations: new buildings were required to have sprinkler systems, and the expanded oil terminals had installed comprehensive foam systems. Fires continued to plague the area, however. In 1968 the National Fire Protection Association declared that Chelsea had more building fires per 1,000 population than anywhere else in North America. Fothergill demanded more equipment, more radios, and better masks and gear. After heated debate, the department scored some victories, including a multimillion-dollar contract to upgrade the city's long-neglected water main system. By 1973 the city had six engine companies, two ladder companies, and about 110 men. Also, the department had purchased an expensive large-diameter hose, and the rank and file were warned to take special care of it. But the fire prevention bureau that had been established in the 1920s had been reduced to a skeleton force of inspectors owing to budget shortfalls. Fothergill—like John Damrell and an earlier generation of fire chiefs—worried about the prospect of a large fire. But the veterans reassured new firefighters: with modern firefighting equipment, there would never be another conflagration like the 1908 blaze. Besides, the rag district would disappear soon; in the early 1970s the area was targeted for urban renewal. The remaining tenements and junk shops were destined to become shopping malls and parking lots—progress in those days.

Sunday afternoon, October 14, 1973, was quiet as usual in the industrial zone. The day was an unseasonably warm sixty-nine degrees. It had been an exceptionally hot autumn and the ground was bone dry. Owing to a depressed market for recycled goods, the bins of rags and papers were overflowing and rubber tires were piling up. A Medford woman who had dropped off her husband at Logan Airport was driving back through Chelsea with her young son when they spotted a fire in the rear of a tin-covered, two-story wood-frame structure at 122 Summer Street, about 200 yards from where the 1908 blaze was first spotted. The former residential structure had been converted into a warehouse, which was now packed with goods. The fire was blazing away and had pretty much engulfed the house. "Let's take a look," the boy pleaded, and the woman, curious herself, stopped the car. She even began snapping pictures. But she grew increasingly alarmed. Surely someone had already called in such a big blaze? So why were no fire engines around? "Come on," she said to her son, and they went looking for a fire alarm box. Nearly three blocks away, at Arlington and Third Streets, she found Box 215 and pulled the alarm, at 3:56 P.M. It was the same box that sounded the first alarm in 1908.

Chelsea Deputy Chief William Coyne, a former Marine and twenty-six-year veteran of the department, was just about to go off duty after returning from a call. It had been a busy day, what with calls about downed wires and other wind emergencies, but nothing serious had happened. When Box 215 was struck, three engine companies and one ladder company responded, Coyne among them. By the time they got to the scene, the fire had already spread and was igniting bales of cloth and paper. Heavy smoke had reached Arlington and Third Streets, about five blocks away. Firefighters first attempted to lay a hose line in front of the blazing building on Second Street, but they were driven back by heat, smoke, and embers. "Strike a second alarm," Coyne ordered. Then, "Skip the second, give me a third." The third alarm brought out Chelsea's two remaining engine companies, plus companies from Boston and Everett, under the region's mutual aid system. Chief Fothergill—C-1 in radio parlance—was working in his backyard when he heard a third alarm ordered via the fire scanner he kept on at all times. He dropped his tools and jumped into his brand-new Chrysler New Yorker. He didn't have to get his chief's car for gear—he had an extra set in his trunk. So he simply sped off toward the fire, stopping two blocks away on Summer Street—a safe distance away, he thought. He arrived on the scene at about 4:05 P.M. and immediately ordered a fourth alarm.

Like sixty-five years earlier, the wind was whipping the blaze into a frenzy and the air was filled with flying sparks. Firefighters set up a line of attack on Third Street, but the wind-whipped fire was overpowering the hose streams, which were weak from a lack of pressure in the area. Companies from other cities began to arrive just minutes after 4 P.M.—from Revere, Everett, Lynn, Boston, and Somerville. The Everett company, arriving at the other end of Summer Street, immediately trained a high-pressure water gun on the blaze, only to find the stream cut off by gusting winds. Fothergill and Coyne ran the length of Spruce Street, trying to get a fix on just how far the fire had spread, but already it was impossible to tell. Coyne kept thinking he had stepped back in time into a World War II bombing scene, with buildings collapsing on either side, sending showers of sparks and bricks into the streets, and with tension wires dropping from poles with hisses and pops. Houses were going up like matches being struck. At 4:28 P.M. Fothergill used the Intercity Mutual Aid radio to report that Chelsea had a conflagration. He asked for all available assistance. (The dispatcher turned from the radio, puzzled that so much apparatus was needed for a—"what was it?"—conflagration. A former Chelsea firefighter, then working in the Mutual Aid office, quickly explained the specific meaning of the term.)

Coyne now could see what some of the men on Third Street could not—that the fire was moving faster than anything he had ever seen. He realized that his companies needed to pull back

Chelsea in flames, 1973

immediately. But the motors of the fire engines had been switched from drive mode to pump mode, and the new hoses were already attached to hydrants and charged with water. There was not even time to uncouple them, even if the men could have manipulated the now red-hot couplings. "Get out, get out," he ordered. "Cut lines and get out of there." The men stared back in amazement. Cut the *new* hose lines? "We've got a good shot at it, sir," one of them shouted. *"Cut the lines. Get out of there,"* Coyne bellowed. Swinging axes, the men sliced through the hoses and pulled out the engines as flames roared over Third Street. The next line of defense was Everett Avenue, a wide thoroughfare that under normal conditions would make for a good line of attack. By now all off-duty firemen were arriving, among them Deputy Chief Bill "Cappy" Capistran. Another former Marine who fought at Guadalcanal, where he had carried a comrade to safety, the short, tough fireman confidently radioed Fothergill, "C-2 to C-1, I'm moving to Everett Avenue. I think we got a shot at it." Fothergill shook his head and said ruefully to Coyne, "Listen to that silly bastard. He doesn't know what's coming." Within seconds, Cappy looked *up* to see fire roaring like a catapult over Everett Avenue. He radioed: "C-2 to C-1, we're moving off Everett." That line of defense was impossible.

Fireman Leo Graves was off duty, but when he heard the third alarm he raced to the scene. Already, thick smoke had covered the area, cutting off sunlight and turning the day into pitch-black night. He ran down Spruce Street, glancing into side streets, noticing with horrified fascination that the fire was jumping streets and moving from building to building. The conditions— eerily replaying those of 1908—couldn't be much worse. Winds were blowing at an average of forty miles per hour with gusts up to fifty miles per hour. By 6 P.M. the blaze had become a firestorm.

The heat was nearly unbearable. Fothergill could see steam rising from the sleeves of his jacket, as if his body were cooking underneath. He saw a firefighter blown off his feet, his comrades struggling to control the wriggling fire hose. Streams of water were being blown sideways, not straight into the flames. More-over, the streams were weak. The fires had triggered the sprinkler systems in the buildings; that water was useless in the over-whelming heat. Even when buildings had been consumed, the sprinkler systems continued to pour out water, decreasing the already low water pressure. The area where the fire started needed at least 4,000 to 5,000 gallons per minute available to have effective streams—but less than 1,000 gallons per minute was available. Michael Drukman, a fire buff and historian, who reached the outskirts of the fire at about 5 P.M., later wrote:

> The flames seemed to have an unusual intensity and we could feel the heat on our faces 300 feet from the flames. The column of smoke was awesome as it rose and curved out over the ocean. The smoke exhibited different colors as it rose, but formed a brown-black cloud as it drifted out to sea.

As in 1908, firefighters needed a wide position from which to form a front against the flames. Fothergill was desperate to get an overall sense of the fire. The state police came to the rescue with a helicopter. Within minutes Fothergill was directing the attack from the air, radioing directions to his deputies below. As flam-ing chunks of lumber soared into the air, he realized that the fire was following the same path as the 1908 fire, up Fourth Street. He feared that the fire station on Everett Avenue was lost, and he could see that City Hall was threatened. But he was determined to keep the fire from spreading to the oil tanks on the waterfront. Perhaps a stand might be made from the area of the Williams School on Arlington Street, between Fourth and Fifth Streets. He attempted to deploy the increasing number of firefighters there.

Meanwhile, Coyne had made an awful discovery on Summer Street. Oh, hell—wasn't that the chief's new car with the fire bearing down on it? Quickly he got the department's mechanic to try to hotwire the car, even as the temperatures were rising to nearly unbearable levels. The mechanic couldn't do it. "I need a sixteen-year-old kid," he yelled in an attempt at dark humor. He and Coyne raced from the scene and Coyne radioed Fothergill: "C-3 to C1, do you have the keys to your car?" "Yep. Right here in my pocket," Fothergill replied. Coyne radioed backed quickly, "C-1, where are you?" "I'm in a helicopter, what do you want?" Wearily came the reply, "Er, C-1, forget it."

Fire companies continued to come, from Malden, Medford, Brookline, Cambridge, Braintree, Arlington, and Lexington. The Tobin Bridge, which had been closed for repairs, was opened for emergency vehicles. By mid-evening, companies were arriving every three to four minutes; in all, about 1,500 firefighters from 95 departments responded. Using the schoolyard next to Chelsea Fire Headquarters as a staging area, Chelsea dispatchers directed the new arrivals to where they were most needed. As Graves noted years later with bleak satisfaction, "There was fire enough for everybody."

As in 1908, nearby residents were leaving their homes, clutching possessions—about 3,500 people were evacuated—but this time the process was orderly. Police Sgt. Andrew Mullen drove through the streets broadcasting warnings in Spanish to ensure that everyone understood the serious situation. Coyne saw someone running with what seemed to be the man's most valuable possession: a portable television. Local clergymen immediately started planning how to house and feed the soon-to-be homeless. Mayor Phil Spelman set up a command post at police headquarters, then moved it to City Hall. At about 10:30 P.M., however, firefighters ordered the building evacuated because sparks were raining on the roof. The slate-covered roof did not ignite, but sparks got under the eves and started burning in the attic. Firefighters managed to douse the flames and save City Hall by climbing on the roof and breaking holes into it to vent the fire; Deputy Chief William Roach had to be carried away by stretcher, a victim of smoke inhalation.

A fierce battle was waged to save the fire station at Everett Avenue. When the heat grew too great outside, firefighters fought from *inside* the structure, using everything, even a garden hose, to keep the flames at bay. A hose line brought in by Boston firemen

Firefighters confer with colleagues during the 1973 conflagration

eventually halted the blaze. As in 1908, the fire held back at the railroad tracks. But the lack of water pressure was frustrating and disheartening to firefighters, who were now exhausted and dehydrated. At one point, Coyne later recalled, Fothergill radioed him to ask, "C-3, anything you need?" Coyne, who almost never lost his sense of humor, said, "Yeah. I'll never get it but I'll say it anyway: Water." "C-3, anything else?" Coyne briefly considered asking if he could begin his promised two-week vacation. "No, C-1."

Graves overheard the radio exchange. He was working with fire companies from Lexington, Abington, and other outlying, even rural, communities to set up a system of relay hoses, a technique employed since the days of the hand engines. One engine pumped water to another engine, which pumped it to another engine—and so on, with as many as five engines forming a link. The "country boys" knew what they were doing; most of them regularly worked without the benefit of nearby hydrants, and they were skilled at setting up a relay. Coyne got his water. He still recalls the satisfaction of placing his foot on the hose and find-

Firefighters in Chelsea, 1973

ing it solid with the precious liquid. But because of the distances traveled by the out-of-town companies and the long hours of pumping, engines were running low on gasoline. One of the more dangerous jobs was performed by auxiliary firefighters, who ran five-gallon jugs of gasoline through the hot streets to fill up tanks.

Back on the ground, an exhausted Fothergill paused with Coyne for a break on the stairs of a synagogue on Elm Street. They recognized the fireman walking up to them: Medford Lt. Frank Cronin. Medford Engine 4 and Ladder were already at the fire when Cronin arrived in the department's much-beloved spare truck, a 1948 Mack 750-gallon-per-minute pumper. The pumper had been dispatched to a house fire on Vale Street a block from the main fire, but the conflagration began moving in fast. Just as Cronin was calling for the men to pull out, the Mack's wooden

Among the losses in the 1973 Chelsea fire was Fire Chief Fothergill's brand-new Chrysler.

steering wheel caught fire. There was no time to cut the lines to back out the truck; all the firefighters ran for their lives as flames overran the truck. Now Cronin was despondent. "What am I going to tell the chief?" Cronin moaned. "I just lost my engine."

"If they're going to kill you for a truck, imagine what they're going to do to me," Fothergill replied by way of comfort. "I have to tell the mayor I lost the city of Chelsea."

But Fothergill hadn't lost the city, not by a long shot. Firefighters at the Williams School were able to make a stand. The school steamed and smoked from the radiating heat but was saved. By 9:30 P.M. the spread of the fire was halted, although the rubble continued to smoke for another four days as companies continued to pour water on the ruins. The "all out" was not sounded until three days later, on October 17. Officially, the cause of the fire was never determined.

The fire had consumed 18 city blocks, leaving 1,100 homeless and another 600 without jobs. It had destroyed about 300 businesses, 60 homes, thousands of feet of fire hose, one Medford fire truck . . . and one 1973 mint-condition Chrysler New Yorker. The cost of the damage was in excess of $100 million. Miraculously, no one was killed, although about 60 firefighters were treated for injuries. Residents walked gingerly through the area, marveling at how it looked like a bombed-out section of London after the Blitz. With lightning speed, the area was declared a federal disaster zone, making residents and businesses eligible for various types of assistance. Rubble was swept away, streets were reconfigured, and urban renewal began. Today a shopping center occupies the area where both conflagrations started.

One afternoon thirty years later, Fothergill, whose son was then the Chelsea deputy fire chief, Coyne, and Graves, all now retired, gathered in Fothergill's kitchen to rehash the 1973 blaze. The fire had become the stuff of legends—and a case study in conflagrations, endlessly analyzed at fire conferences and conventions. (Just a week after the fire, for example, Fothergill and Coyne showed footage of the fire at a meeting of the International Association of Fire Chiefs in Baltimore.) They swapped tales about Cappy, now gone, and the former chief joked for the hundredth time about his new car and how he came close to "losing" the city. Fothergill admitted that he had often become depressed discussing the 1973 fire at conferences, particularly when he had to answer questions about what was lost. Then one day a friend and NFPA official suggested that instead he emphasize what was saved, like the Everett Avenue fire station, the Williams School, and City Hall. The entire business district was spared. Although the fire started under nearly the exact same conditions and in almost the same spot—in an area still begging for a burning—the 1973 conflagration was only a third the size of the 1908 blaze. And not one life was lost. "You didn't lose the city, pal," Graves told his former chief, and Fothergill had to agree.

7

THE COCOANUT GROVE

Heat, Smoke, and Panic

There has never again been—and firefighters most fervently hope that there never again will be—a fire like the one in Boston's Cocoanut Grove nightclub on November 28, 1942. Nothing about that fire was normal—not its terrible speed, not its mysterious fumes, not its strange twists of fate that left some without a scratch and others to die horrible deaths. Unlike the 1872 fire described in Chapter 5, which raged for two days and took relatively few lives, the Cocoanut Grove blaze was over in less than an hour and killed hundreds. It was the worst death toll by fire in a public place since 600 lives were lost at the Iroquois Theater in Chicago on December 30, 1903. Few Bostonians were unaffected; if they didn't have a friend or relative at the club, they knew someone who *almost* went there or left because it was too crowded. If they didn't have a firefighter in the family or a relative among the city's medical staffs, they knew of a mother in labor who couldn't get a hospital room or of a funeral delayed because no coffins were available. Even today the Boston Fire Department continually fields requests for documents on the blaze. Fire buffs still debate its true cause; lawyers, doctors, and building contractors continue to work under its impact. For most Bostonians, however, the true story of the Cocoanut Grove is its devastating effect on the hundreds of individuals who were caught in a maelstrom of heat, smoke, panic, and pain—individuals out for a night of fun who were doomed by greed and thoughtlessness.

In the uneasy years during World War II, Bostonians grabbed eagerly at any bit of joy they could wring from a difficult life. In the morning they would sadly peruse the city's newspapers for the names and photos of New Englanders who had died in Europe or the Pacific. After work they would flock to the movies to see stars like Bob Hope and Bing Crosby in *Road to Morocco* and Clark Gable and Lana Turner in *Somewhere I'll Find You*. On Saturday night they would pack the city's nightclubs—the Latin Quarter, the Mayfair, and the Cocoanut Grove—and toast one another with a post-Prohibition abandon. Moreover, they rallied around their sports teams. This being Boston—now a very Irish, very Catholic Boston—they rooted with special vigor for the Boston College football team.

When they awoke on Saturday morning, November 28, 1942, Bostonians believed they would have something to really celebrate by evening. Boston College's undefeated football team was expected to walk over the lackluster Holy Cross squad in Fenway Park that afternoon. Practically assured of a trip to the Sugar Bowl, Boston College (BC) officials had already booked a victory party at the Cocoanut Grove for that evening. In the city's swellest and most notorious nightspot, they would toast their victory.

But the football gods are fickle. BC went down to defeat 55–12, before a disbelieving and devastated crowd. Mournful BC officials canceled their nightclub party. Still, many BC fans went to the club anyway. After all, the Cocoanut Grove was the place to be in Boston on a Saturday night, and if they were going to drown their sorrows, they might as well do it in style. As for the Worcester fans who had come to Boston to root for Holy Cross, they would toast the upset with real pizzazz.

The football disaster notwithstanding, life went on as usual around Boston that afternoon. Veteran firefighter Charles Kenney, for example, had other things on his mind. He and other firefighters were still mourning the loss of six fellow firefighters who had died just two weeks earlier in a fire in the Luongo Restaurant in East Boston.

By now the fire department had fifty-three engine companies, thirty-one ladder companies, three rescue companies, three water towers, and three fireboats. Like all the city's firefighters, the 42-year-old Kenney wondered if some day he, too, would be asked to make the ultimate sacrifice. As he walked from his home in the South End to the Rescue 1 station on Broadway near Washington Street, he passed, as usual, close to the Cocoanut Grove, bordered by Piedmont, Shawmut, and Broadway Streets on the edge of the city's theater and nightclub district. He hardly gave the club a thought.

In East Boston, in the Engine 9 company, John F. Crowley was hoping for a chance to prove himself to the more experienced firefighters. Recently hired, he had worked two-day shifts, and now he would start his first twenty-four-hour rotation. His $33 weekly salary was important for his growing family, but Crowley wanted more than the money. The 27-year-old already sensed that he was on the edge of a special community, a kind of brotherhood, bound by honor and dedication. His uncle, Daniel Crowley, was a district chief, but John wanted to win the respect of that group on his own merits, and he knew it wouldn't be easy. Tonight he would learn the running cards, the forms that stipulated which fire companies and what apparatus were to respond to successive fire alarms. As he was puzzling over the notations, Box 1521 signaled a first alarm at 10:20 P.M. Then, almost before the first alarm registered, another alarm from the same box sounded—a third alarm. Bobbie Quirk, of Ladder 2, leaped out of his chair at the sound. "What the hell is going on?" he snapped. "They *never* skip an alarm." Engine 9 would respond, to cover the station of Engine 7, on East Street, near South Station. The men ran for the engine with an excited but nervous new guy along with them. They arrived at East Street within minutes but were immediately dispatched to the scene itself. The Cocoanut Grove was on fire, and almost every fire company in the city was at or racing to the spot.

The call would be John Crowley's very first fire. It would be Charles Kenney's last.

How did a club like the Cocoanut Grove—with a name redolent of warm beaches and swaying palm trees—ever wind up in cold, staid Boston? How did a nightspot with a notorious past attract Boston College officials and middle-aged couples as well as sailors out on the town and wise guys on the prowl? The answer goes back to the 1920s and a talented kid from Roxbury—a furniture salesman with showmanship in his blood and bright lights in his eyes. He was born Milton Irving Alpert in 1904, but he was always known as Mickey. As a teenager he launched a singing career warbling radio ads for his furniture company and won plaudits for his roles in local theater productions. With his curly hair, dapper style, and quick wit, he was a natural comedian and showman for the vaudeville era. He needed only a stage to turn on the magic charm. And he found a likely building to house that stage at 17 Piedmont Street. It was an unassuming structure that had previously housed a garage and a film distribution company. He also found a mysterious fairy godfather in Jack Berman, a wealthy California businessman who had become friendly with Mickey and his friend Jacques Renard, a talented bandleader. Berman had a proposition: Together they'd build a nightclub. He would provide the money, Alpert and Renard would provide the entertainment and run the place. Alpert pulled his brother George, then the youngest ever first assistant district attorney for Suffolk County, into the deal. George, however, insisted that the club strictly adhere to liquor laws; setups were fine, actual booze was not. Prohibition was the law, ratified by the Eighteenth Amendment in 1919, and George didn't want the family name sullied by a speakeasy raid. What the club lacked in booze, it would make up for in glamour. At Berman's insistence, famed night-club architect Reuben Bodenhorn was hired. He created a tropical paradise in chilly Boston: he dotted the dining and dancing room with fake palm trees, covered the walls in leatherette and rattan, and designed furniture with a stark zebra pattern. The ceiling was painted to suggest a night sky, and on a hot summer night the roof could be rolled back for dancing under real stars. Alpert and Renard hired as maître d' Angelo Lippi, known as "the Count," whose tiny mustache and ever-unruffled demeanor added another element of elegance. Berman reportedly spent $85,000 on the remodeling; he even suggested the name "Cocoanut Grove," after the similarly named club in his home base of California. Renard and Alpert were too enthralled by the cash flow to be suspicious. "Maybe he prints his own money," Renard mused to Alpert. "Maybe he's got a gold mine. I'll say one thing, there's nothing cheap about the bloke." Strangely, Berman wanted to stay in the background, leaving Alpert and Renard to take front stage. Or perhaps it was not so strange, for Jack Berman turned out to be Jack Bennett, a con artist with the feds on his tail. Just before the club was to open, Alpert and Renard learned to their horror that Berman/Bennett, a partner in the Julian Petroleum Company, was wanted for manipulating the oil stock market. The fairy godfather proved more godfather than fairy.

Horrified, the pair called in George, who used his legal finesse to work out a deal to buy the club. Alpert and Renard scraped up the money, and in a blitz of extravagance, the club opened on October 27, 1927. For a couple of years, Alpert and Renard thought they would make it, with Renard as the excellent bandmaster and Alpert as the irrepressible master of ceremonies. But under the grip of Prohibition, and after the stock market crash of 1929, people stopped coming and the money stopped flowing. Without booze and Berman, the club couldn't stay afloat. Moreover, Alpert wanted out; he wanted to go to New York to launch a career there.

Waiting in the wings was Charles "King" Solomon, a thug with a flair. A heavyset, round-faced mobster, Solomon ran liquor and dope smuggling and other businesses spoken about only in whispers. With his connections to the Murder Incorporated wise guys in New York City, he was untouchable in Boston, "short of a particularly messy murder," as *Record American* reporter Austen Lake declared. Through his lawyer Barnett Welansky, Solomon made an offer to the Alperts that they decided they could not refuse. For $10,000 the club passed into Solomon's hands in 1931. Renard tried to put up a legal fight, but his stock in the club was judged worthless and the sulking musician decamped to a rival nightclub.

While middlemen and fake fronts shrouded many of his holdings, Solomon used the Grove as his personal showcase. True, the joint was perfect as an "office" for making deals, but here the King could rub shoulders with stars, starlets, and fawning officials. He brought in the best of vaudeville: Texas Guinan and her "Hello, Suckers" comedy review; Sophie Tucker; Gilda Gray, the original "Hula Girl." The pretty girls were always seated at Solomon's table. Other stars who lit up the club were Jimmy Durante, Rudy Vallee, and Guy Lombardo. Solomon retained Angelo Lippi as the maître d', even though (as Lippi later said) Solomon could be slow with a paycheck. But one didn't bother the King with something as trivial as one's daily wages.

Of course, Solomon winked at Prohibition, but the ill-conceived law was repealed in December 1933. The night the first legal drink was poured was one of the most memorable at the Grove; indeed, workmen were still hammering nails into a hastily constructed bar as the first champagne corks were popped. Asked to make a speech, the ever-suave Lippi put it simply: "My dear ladies and gentlemen, it is my sublime pleasure to inform you: the bar is open."

The King's dark side always simmered underneath his munificent, star-struck surface. He made sure the club's exit doors were locked on the inside to prevent anyone from skipping out on a bill. Limited access also protected him against would-be hit men. His paranoia was not unjustified. In January 1933, at the Cotton Club in Roxbury, a group of men hustled him into the bathroom and shots rang out. His assailants fled and Solomon staggered out clutching his gut. "The dirty rats got me," he croaked before collapsing. Not surprising, the fallen King was nearly "bankrupt" (front men know when to stay quiet). The Cocoanut Grove was listed among his assets as having "no value." One of the lingering mysteries is how quickly ownership of the club passed from Solomon's widow, Bertha, to Barnett Welansky, "the bland, monosyllabic young lawyer," as Lake dubbed him. Despite his association with Solomon, Welansky was a surprisingly well-educated and on-the-level attorney. Born in Boston in 1896, one of six children, he grew up an ambitious, energetic kid with a yearning to be a lawyer. He sold papers to earn money for college and eventually got his law degree in 1918 from Boston University and his master's degree a year later. After being admitted to the bar in 1919, he joined the prestigious practice of Herbert F. Callahan. He was on his way to being a "typical American success story," as the papers later put it, until he started representing Solomon.

Welansky's younger brother, James, was another matter. James was a player in the city's nightclub scene. He was managing Boston's Metropolitan Hotel when the notorious racketeer and gambler David "Beano" Breen was shot to death in the hotel lobby in December 1937. James promptly skipped town, only to resurface in Florida two months later, claiming that he just decided to take a long-planned vacation . . . without luggage and under an assumed name. Arrested for the murder by prosecutors who claimed he and Breen had run a gambling operation, James Welansky insisted that he did nothing and saw nothing. A grand jury refused to indict him. So he went to work for his brother and took charge of the Cocoanut Grove when his brother wasn't around.

Barnett Welansky ran the Grove like a business, not a showcase. He bought adjoining property and expanded the club; booking stars was secondary to making profits from the now free-flowing liquor. In 1938 he built the downstairs Melody Lounge, a dimly lit piano bar that retained a speakeasy atmosphere. In early November of 1942, Welansky opened a new cocktail lounge, accessible to customers only by a door on Broadway, although staff could enter via a passageway from the dining room.

He wooed back Angelo Lippi to manage the place and installed Rose Gnecco Ponzi to keep the books. (Rose's ex-husband, Charles Ponzi, had the dubious distinction of creating the financial scam that bears his name: the Ponzi scheme.) More important, Welansky

Cocoanut Grove souvenirs

brought back the man long associated with the Grove: Mickey Alpert. New York hadn't been the success Alpert had dreamed it would be. So he came back to the scene of his former dream to act as master of ceremonies. Other regular entertainers were lined up, including singer Billy Payne and pianist/singer Goody Goodelle. Solomon's showplace became Welansky's cash cow.

But Welansky suffered from Solomon's disease: he worried about patrons skipping out on their tab. His head cashier, Katherine Swett, lived in fear of his wrath—and that fear would prove fatal for her. Being a club owner also caused Welansky to relax his scruples. Unlicensed electricians worked on the club's lights and wiring—why bother to get the legal paperwork when, as a contractor later testified, Welansky implied that Boston Mayor Maurice Tobin owed him favors. "Mayor Tobin and I fit," he bragged.

The club was now a complex: Downstairs was the Melody Lounge, appointed with a circular bar, a piano stand, and fake palm trees. Folds of fabric were suspended about eighteen inches below a concrete ceiling to create a sense of intimacy. At ground level was the main dancing and dining room, with a rolling stage for the band and a raised terrace for important guests and celebrities. Just off the dining area was the Caricature Bar, lined with drawings of the well-known. The bar itself, at forty-eight feet, was the longest in Boston. Dressing rooms for the chorus girls were on the second floor, and a large kitchen with a walk-in refrigerator was in the basement. The new Broadway Lounge featured a spiffy modern decor, with a row of the popular glass-block windows. The legal occupancy limit for the entire club was set at 600.

Customers had only one entrance into the main part of the club: They passed under an archway on Piedmont Street and through a revolving door. To get into the new lounge, customers passed through an outside door on Broadway into a small vestibule and then through another door. An emergency exit, equipped with a panic bar, was located down a small hallway at the top of the stairs to the Melody Lounge; a double door on the Shawmut Street side was hidden by curtains. There were service entrances in the basement and behind the orchestra stage.

Despite the war, business never looked better in 1942. Prohibition and the Great Depression seemed like a dream. Customers applauded Billy Payne, laughed at Alpert, and downed whiskey sours and shots of bourbon. They ate half a dozen oysters for 40 cents, followed by a sirloin steak for $1.90 or a baked lobster for $2.25. On November 20 the club was inspected by Lieutenant Frank Linney of the Fire Prevention Division, who declared the club's condition "good," finding enough exits and no flammable decorations.

Welansky's kingdom was complete—until the night of November 28, 1942, when in less than half an hour, the faux paradise was swept away with hundreds of lives.

INTO THE FIRE

Everyone in Boston knew Marty Sheridan, the reporter with the sharp wit, thick glasses, and cocky charm. A 28-year-old freelancer

for some of the city's nine papers, Sheridan knew the city's play-
ers, both high society and low life. A native of Rhode Island who
moved to Boston to pursue a writing career, Sheridan had published some short stories, plus a well-received book on the history
of comics. Now he mostly wrote features because, as he jauntily
wrote in his book's bio, "there is more and quicker money in it."
Sheridan had a knack for writing personality features and boasted,
"The bigger the person is, I find, the easier he is to handle." Even
though he hoped to write another book soon, freelance writing
was not always lucrative, and he had a wife to support. So he did
public relations work part-time. He'd escort famous people around
Boston, interview them on radio broadcasts, and snap their pictures. Since he enjoyed meeting people, he didn't mind the work.
Once he even escorted the popular young comedian Bob Hope. In
November 1942 Monogram Pictures asked Sheridan to escort
another major celebrity: Buck Jones.

Any kid in Boston could tell you all about Buck Jones, the famous
cowboy star. Born Charles Gebhard, in Indiana, Jones grew up rid-
ing horses in Oklahoma. After a stint with a circus, he started
doing stunts for Hollywood westerns, churned out conveyor-belt

*Martin Sheridan (right) interviewing Buck Jones just hours before the
Cocoanut Grove fire*

fashion. By the early 1920s the six-foot Jones, with his rugged build and laconic charm, had ridden into the corral of top cowboy stars like Tom Mix and William S. Hart. By 1939 he had starred in nearly 120 movies. Fan club members—the Buck Jones Rangers—numbered in the millions, and his fan mail reached as high as the Rockies. In the late 1930s, however, his appeal began to wane, as his tough loner image lost ground to those "sissy" singing cowboys like Gene Autry. But Buck, now the owner of a California spread and a much-beloved boat, continued to find work, even if he sometimes (the horror!) had to play a bad guy. Movie producer Scott Dunlap connected him with Monogram Pictures, which specialized in low-budget films like the Charlie Chan series and the Bowery Boys, for its "Rough Riders" series. Monogram was anxious to promote Buck's latest films, *Dawn on the Great Divide* and *West of the Law*. So Jones was coming to Boston for the last few days of a two-week national tour to promote war bonds and to drum up interest in his new movies.

Sheridan picked up Jones and Dunlap at Boston's South Station on the night of November 27; Jones was wearing his cowboy hat and signing autographs for some servicemen. Sheridan had meticulously planned Jones's Saturday schedule down to the minute: a visit with ailing kids at Children's Hospital at 9:30 A.M.; Junior Commando rally at Boston Garden at 11 A.M.; and cocktails and luncheon with theater people and the press at the Statler Hotel at 12:15 P.M. Jones would also see part of the BC–Holy Cross football game from Mayor Tobin's box, do a radio interview, and last, appear at the Buddies Club. Sheridan even arranged for the loan of a horse from a Boston police officer, and he managed to locate a Samson Spot twenty-foot lariat. Then plans shifted a bit; in the early evening, Jones taxied to Newton for a cocktail party with local movie executives. Because he was suffering from a cold, he begged off going to the Buddies Club. He wanted to go back to his hotel and rest. Sheridan obligingly canceled the appearance. But the local movie bigwigs weren't about to let the cowboy ride off into the sunset without having a bit of his star power rub off on them first. They had planned a party at the Cocoanut Grove and insisted that he come along. Jones, ever the stalwart cowboy, felt he couldn't let them down. Sheridan, on the other hand, was grumpy. He had never been to the Cocoanut Grove and had no desire to go there. Still, he picked up his wife, Connie, who was happy about showing off her new mink coat, and joined the party.

Seventeen-year-old Dotty Myles was star struck, no doubt about it. Funny thing was, everyone around her was equally certain that her dreams would come true. Blessed with a lilting voice, luxuriant hair, and a face as fair and fresh as a flower, she was also an accomplished pianist and had the stage presence of a natural actress—she was particularly good at mimicking accents. She was born Dorothy Metzger, in New York City, under the shadow of Broadway on West 101st Street. By age 9, she was already taking voice lessons and by 12 was winning voice contests. In her late teens she started singing in nightclubs, and in December 1942, at only age 17, she was slated to audition for Jimmy Dorsey himself. First, however, she wanted to get warmed up. What better way than a three-week gig at Boston's famed Cocoanut Grove? When she debuted on Sunday, November 8, the enthusiasm of the crowd was a sign. Her singing career was about to take off.

That Saturday the town was filled with servicemen and sailors. Many of them wanted to take dates to the city's most impressive club, the Cocoanut Grove. John and Claudia O'Neill went there to celebrate their marriage, just a few hours old, with members of the wedding party. Joseph Dreyfus, a medical student and hardly a regular nightclub goer, was there with his wife and another couple at a send-off party for an acquaintance who was going overseas. Twenty-one-year-old John Quinn, of Worcester, was ready for a night on the town: he had just two more days before reporting back for duty at the U.S. Naval Training Center in Newport, Rhode Island. Quinn and his buddy Dick Vient were thrilled by the Holy Cross victory. So they had arranged to meet their girls and get in some partying before leaving for the Pacific. Quinn was bringing his gal, Gerry Whitehead, a girl he had loved since his junior year in high school. He knew she was the one and that they would get married after he returned from the war. Vient brought his fiancée, Marion Luby, and the foursome managed to get seated at a choice table near the dance floor. When Quinn saw his favorite cowboy star, Buck Jones, right there in the club, he couldn't resist saying, "Hi, Buck." "Hi, sailor," the big man replied, and Quinn thought that was just swell.

Sheridan, on the other hand, was unimpressed with the joint. Too crowded, too noisy, too smoky. His party—about thirty people in all—was crowded around two tables on the raised terrace, the section reserved for celebrities. It was so jammed that guests were passing drinks and dishes overhead for the waiters. Sheridan ordered an oyster cocktail and hoped the night would pass quickly. He was pressed up against the wall, and when he casually put his hand against it, it felt curiously warm.

By 10 P.M. more than 1,000 people had jammed the Cocoanut Grove, perhaps twice the legal occupancy limit. The coatroom was completely filled, and fur jackets were piled on the floor. It was so crowded that a number of disappointed patrons left for other clubs—not realizing that the fates had smiled on them. Waiters were rushing food from the kitchen to the dining room above. Club owner Barnett Welansky was not there; he was in the hospital because of a heart condition.

In the main dining room, Mickey Alpert was waiting for the right moment to begin the second show of the night. In the downstairs Melody Lounge, Goody Goodelle was performing songs aimed at tickling the servicemen clustered at the bar. The venerable Lippi was nursing his arthritis at home, leaving wine steward Jacob Goldfine in charge. In the new Broadway Lounge, bartender Tiny Shea—so nicknamed because he topped the scales at 385 pounds—was mixing drinks. James Welansky was drinking with Boston Police Night Captain James Buccigross, who was supposedly making his rounds, and Suffolk County Assistant District Attorney Garrett Byrne.

A little before 10 P.M., on the way to the men's room, Quinn and Vient noticed the stairway to the Melody Lounge and took a peek. The lounge looked so cozy, with the singer on a revolving piano stand crooning "Never Let a Sailor Get an Inch above Your Knee," that they decided to get the girls and go down.

And just about that time, a man in the Melody Lounge decided he needed a little more intimacy with his honey and reached up and unscrewed a light bulb in the fake palm tree above his head. Bartender John Bradley immediately noticed; "Hey buddy, you can't do that," he sang out. The man ignored him. So Bradley got the attention of a busboy. "Stanley, go out there and put that bulb back in." Stanley Tomaszewski, working illegally in the club at the age of 16, jumped on a table and reached into the fake palm fronds for the bulb. It came away in his hand. Now he couldn't see the socket. So he reached in his pocket for a Cocoanut Grove matchbook, lit a match, and located the socket. He put out the match and screwed the bulb back in. Almost immediately, a patron saw a tiny flare or spark in the ceiling. Then flames started nibbling the fake palm tree. For a second it seemed funny—"Hey, look, the palm tree's on fire"—but only for a second. Tomaszewski tried to beat out the flame with his hands. Bartenders ran over with water. But the fire started to spread through the fabric suspended from wood strapping under the concrete ceiling. Within a minute, no one was laughing. The Melody Lounge was on fire.

Some customers, the lucky ones, began walking up the stairs toward the revolving doors. But the fire was spreading fast. Bradley and Tomaszewski started yelling for people to follow them into the kitchen and to exits to the outside. But with a speed that defied comprehension, the lounge filled with flames. Panicking patrons jammed the staircase, the fire roaring behind them. Many rushed to the Piedmont Street exit door, the one with the panic bar. They pressed on the door. It didn't budge. Fruitlessly the growing crowd beat on the door, trying with all their strength to break it down. Some patrons managed to get into the foyer leading to the main dining room, intending to get through out the revolving doors. Some foolishly stopped to retrieve coats. A waiter—even more foolishly—is reported to have yelled, "No one leaves until they pay their bill." A few managed to slip out the main entrance. But within seconds people, now with their hair on fire and skin blistering, had jammed both sides of the revolving door, effectively rendering it impassable.

Behind the bar in the Melody Lounge was 24-year-old Daniel Weiss, a medical student at Boston University and Welansky's nephew. Weiss had hesitated when the fire broke out, hating to leave his post at the cash register. Just as he was making up his mind to leave, the lights went out. He dropped to the floor coughing. He managed to soak a bar towel with water and wrap it around his nose and mouth. Lying on the floor, he found it easier to breathe, while the screams of other patrons grew louder and louder and then faded away.

John Quinn and his companions had barely made it down the stairs into the Melody Lounge when one side of the room lit up; Quinn could see that one of the fake palm trees was on fire. Quickly, but without panicking, the foursome turned around and started up the stairs, passing what appeared to be a waiter with a fire extinguisher. So it would be over soon, Quinn figured. Almost instantly the man rushed back past them. The foursome reached the foyer of the main dining room and a crowd of increasingly nervous customers. Quinn felt a blast of heat sear the back of his neck. Turning, he saw a sheet of bright orange flames stretching from ceiling to floor and from wall to wall. He didn't look back again. Ahead of him Gerry, Dick, and Marion seemed to be headed for the coat check. Quinn knew that the main entrance was just to his right, but he sensed that people were already backing up behind the revolving doors. And then the lights went out.

The narrow staircase from the Melody Lounge had acted as a chimney, sending heat and carbon monoxide upward. The fire moved so fast that the Melody Lounge was consumed before people in the main dining room were even aware of any trouble. Seated on the terrace, Sheridan heard a commotion coming from the foyer. To him it sounded like "Fight, Fight." Oh, great, he thought, a brawl between some Boston College and Holy Cross fans. Then he saw smoke and heard the cry again: *"Fire!"* He could hear flames crackling but didn't see any. "Let's get out of here," he said to his wife, keeping his voice casual. "How about my mink coat?" Connie cried. "To hell with it. Let's take our time and get out of here." They rose, took a single step, and the lights went out. Within seconds they were gagging on fumes and smoke that engulfed the entire room. As if clubbed with an invisible bat, Sheridan passed out, the shrieks and sounds of breaking dishes and glasses growing ever fainter. He and Connie sank into a growing pile of bodies.

Most patrons didn't know about the double door, hidden by draperies, that led to Shawmut Street. The wall also had three plate-glass windows, but they were covered by wood veneer. A waiter, alerted to the growing fire, pulled aside the drapes and pushed at the door. It, like so many other exits that evening, was locked. Men rushed to help him, including, coincidentally enough, John Walsh, Boston's civil defense director. They managed to open one of the double doors, and people began pouring out.

The dining room had turned into bedlam. It wasn't a matter of smelling smoke and getting concerned. It wasn't as if the ceiling, then the walls, then the furniture caught fire. All of a sudden a ball of flame, described variously as bright orange or bluish with a yellow cast or bright white, roared through the dining room, rolling along the ceiling like a fiery tsunami and filling the air with fire before anything had begun to burn. When the lights went out, the now terrified patrons hardly knew where to turn. To the only exit they knew, the revolving doors? Or away from the flames toward the Shawmut Street side? Meanwhile, they were gagging on the horrendous smoke—thick, acrid, and laced with a strange, sweet smell that stuck in the nose and throat.

Quinn had made a quick choice. Deciding that the revolving doors would quickly jam, he headed for the Shawmut Street door. He would push Gerry into the middle of the crowd and hope that they would be squeezed like toothpaste from a tube, to the other side. Grabbing Gerry by the waist, he whispered, "Do exactly as I tell you and don't say anything until we get out." Gerry broke that promise only once. They forged ahead into the mob as black smoke filled the room. They would take a step, then Gerry would trip over something, maybe a body, but Quinn did not dare to look down. He would bend to lift her to her feet and, rising, lift as well those people who had climbed onto his back. He saw two men in tuxedos trying to swim out by pulling the hair of people in front of them. He felt his knees buckling. Falling to one knee, he put a hand on Gerry's back, determined to push her out with his last strength. It was then that she broke the code of silence. "Air," she gasped. Galvanized by that one word, Quinn pushed himself to his feet and, squeezed on both sides, he and Gerry were pushed out the door. They landed on the hood of a car without touching the sidewalk. The tube-of-toothpaste strategy had worked.

Dotty Myles had arrived early on Saturday evening, about 9:15, and thought she'd use a few moments to study algebra before she went onstage. At about 10 P.M. she saw a strange glow coming from the Melody Lounge and realized that the club was on fire. She could see the door on the Shawmut Street side and tried to cross the dining room to get to it. Someone knocked her to the ground, and before she could get up, an overturned table hit her squarely in the face. She blacked out and came to with the pressure of feet on her body, the heavy tread of men and the stabbing impact of high-heeled shoes. She heard moans, shrieks, and someone calling, "Mother . . . Mother . . . Mother." She herself was praying as she reached up and touched a man's hand and found herself yanked to her feet, her gown completely torn away. That was a blessing: she could see women, as bright as torches, as their evening gowns burst into flames. She tried to follow the man who pulled her up, but he plunged ahead into the swirling mass of people. She tried to push forward but fell into a sea of people that piled on top of her. She could feel the skin on her arms burning as if it had been splashed with acid, and then she didn't feel anything at all.

Panic is not merely an emotion; it is a state of being. As smoke, fumes, and then a wall of roaring flames filled the dining room, humans ceased to be human. Simply, they were stripped of the ability to reason and act rationally. Think of how fast you react, without thinking, if you accidentally touch a hot stove. The heat in the Cocoanut Grove reached that kind of temperature. Most of the people reacted with the same raw instinct, the primal need to flee—even if they didn't know where to go. Those that could move pressed toward the Shawmut Street exit. So, unfortunately, did the fire. Flames, which seek oxygen to burn, quickly consumed all the oxygen in the dining room and rushed toward the doors opening onto Shawmut Street. The exit was transformed from a conduit to safety into a blowtorch.

In the Broadway Lounge the drinking continued—until a hostess rushed up to James Welansky to tell him that the club was on fire. Barely had the words left her mouth when smoke started to pour into the lounge, followed by a rush of people who had managed to navigate the passageway from the dining room. Captain Buccigross attempted to call for order; he was knocked aside. Somehow, though, both he and James Welansky managed to get out. Others were not so lucky. Of the two doors into the lounge, an inside vestibule door opened inward, against the flow of human traffic. The crowd pushed the door shut and the crush kept it shut. Like the revolving doors, like the stairs from the Melody Lounge, the Broadway Lounge became a death trap.

In the dining room, club waiters and other staff attempted to direct people out the doors behind the stage or through the basement to other exits. But those who pounded on the service doors behind the stage leading to Shawmut Street found them locked. While some managed to get into the basement and squeeze through windows, others were trapped in the crush. Saxophonist Al Willet was among them. He and about forty others couldn't move, and they began to choke on the terrible smoke. Willet pulled out a handkerchief and pressed it to his mouth. "I guess this is it," he said to bassist Jack Lesberg, before they both sagged to the floor.

Fire kills in many ways, and the club victims got the full range. Some were burned alive or died later from their burns. Others suffocated as the fire robbed their lungs of oxygen. Still others

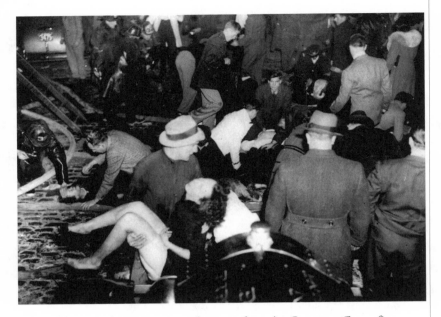

One of the most famous images of rescues from the Cocoanut Grove fire

were poisoned by carbon monoxide, the gas produced by incomplete combustion. Some who managed to get out of the club seemingly uninjured simply collapsed in the freezing air. Others were felled by an irritating gas or by breathing superheated air, which made their lungs fill up with fluid; they drowned in their own secretions, victims of pulmonary edema.

The fire's speed and ferocity still baffles fire analysts. In a summary often repeated in histories of the fire, Paul Benzaquin, in his 1959 book on the fire, declared, "Twelve minutes after the tree caught fire, everyone who was to die was dead or mortally burned." Flames broke out in the Melody Lounge about at 10:15 P.M. Within a minute the fire had spread to the foyer upstairs. By 10:18, the main dining area was embroiled. Two minutes later fire raced through the passageway into the Broadway Lounge. By 10:35 P.M. the first victim arrived by police car at Boston City Hospital. And by 10:45 P.M. firefighters had the main body of the fire under control. But, strange to say, the fire could have been

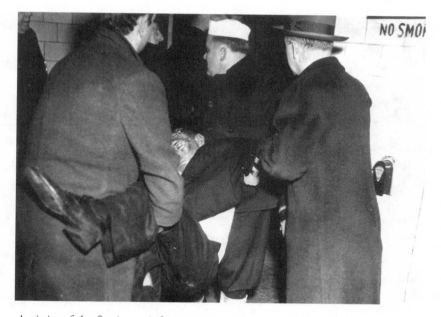

A victim of the fire is carried out past a NO SMOKING sign

worse. Of the strange coincidences of the fire, none was stranger that night than the minor fire that brought firefighters within blocks of the Cocoanut Grove just as the Melody Lounge was about to go up in flames.

At 10:15 P.M. a citizen pulled box 1514; a fire had broken out in a car near the corner of Stuart and Carver Streets. Because the box was in a high-density area, a large contingent of men and apparatus responded: four engine companies (Nos. 35, 10, 7, and 22), two ladder companies (Nos. 13 and 17), and Rescue Company 1, with Deputy Chief Louis Stickel and District Chief Daniel Crowley. Engine 22, first on the scene, extinguished the car fire within a minute; it turned out to be just a minor blaze. Lt. Miles Murphy and Charles Kenney of Rescue 1, Captain Jeremiah Cronin and George "Red" Graney of Engine 35, and other men were chatting leisurely, putting away equipment, and getting ready to return to the station, when several of them heard a commotion and thought a fight—probably between a couple of

Crowds outside the Cocoanut Grove on the night of the fire

sailors—had broken out at the nearby Cocoanut Grove. Then the shout rang out, "Hey, there's another fire," and the firefighters could see trails of smoke. Almost at that instant someone pulled Box 1521, and the wife of an off-duty fireman, who had already rushed downstairs, was telephoning to report that she could see smoke at the Cocoanut Grove from her second-story window. Members of Engine Company 22 leaped back into their engine and headed up Broadway to Shawmut Street. Pulling up, they could see people running from the club, hear screaming, and then, when they reached the corner, flames were roaring out of the door of the Broadway Lounge. Jumping from their rig, the men started dragging a hose toward the Broadway door to get water into the new lounge. Around the corner Kenney could see flames pouring from a door on Shawmut Street. People with their hair and clothes on fire were stumbling from the door; more lay still on the ground.

As Lieutenant Murphy focused on the Broadway side of the club, Kenney dashed toward the inferno at the Shawmut Street door. He could see what looked like a growing pile of bodies inside. "Get a line to play in the door," Kenney yelled. He didn't wait for the water; he plunged into the building, desperate to get some of the people out. Meanwhile, as water played over the

Broadway door, Ladder 13 firefighters swung axes at the glass-block windows, struggling to gain access. But the modern windows resisted the fiercest blows. When the men finally broke through, they could see blackened hands reaching for help that was coming too late.

Told by District Chief Crowley that people were trapped inside, Deputy Chief Louis Stickel skipped the second alarm and ordered a third; it went out at 10:23. A fourth alarm followed within the minute.* Other engine companies, alerted by the first Box 1521 alarm, screeched up to the Piedmont Street side of the club, momentarily unaware of the drama around the corner. Fire was shooting twenty feet out from the main entrance. Peering through the arches, firefighters could see the panicked pileup at the revolving doors. When they finally smashed down the revolving doors, they were driven back by intense heat and flames. All they could do was pour water into the entryway and watch helplessly as people burned to death. Nearby, firefighters tried to break down the door that led to the foyer and the Melody Lounge. It would not open. Quickly bringing a battering ram, they smashed the door open. Inside, all they could see were dozens of bodies. Over on Shawmut Street, Graney had managed to get a charged, high-pressure hose inside the main dining room. In a chaos of flame and water, Kenney and other firefighters were pulling people and bodies out the door, their hands blackened by heat from the hot flesh of victims. Firemen smashed the windows on the Shawmut Street side and broke through the thin plywood interior, trying to vent the fire and provide more exits. Working on pure adrenaline, Kenney tried to

---

*Various accounts disagree over which fire officials ordered which alarms. Even the official records are confusing. In his report, Stickel said he ordered the third and fourth alarms, based on Crowley's assessment. District Chief William Mahoney said in his report that when Deputy Chief John McDonough ordered him to send a second alarm, he learned a fourth alarm had already sounded. Even though the main body of the fire was out by that time, a fifth alarm was ordered at 11:02 P.M., either by Fire Chief Samuel Pope or by Deputy Chief Mahoney as a way to bring other companies to relieve the firefighters "who were showing exhaustion from the gruesome tasks," as Mahoney's statement put it.

reach more people inside the Shawmut Street doorway, coughing as the smoke filled his lungs. Then he saw, in a pile of bodies, a woman's small hand desperately waving. "Hold on, sister, hold on," he cried and firmly grabbed her wrist, even though he could feel his fingers sinking down through burned flesh to bone. "Take it easy. We'll get you out of there in a minute." He could feel her hand grabbing his, grabbing with all her strength. He managed to pull the woman—she was no more than a girl—to her feet. Her clothes were torn away and she was badly burned, but Dotty Myles was alive. Kenney carried her to one of the servicemen now flooding the scene to help and dashed back into the club to pull out more people and bodies. And then he couldn't breathe—the smoke, the fumes, and the heat had seared his lungs. His knees buckled, and in a second he had fallen to the ground. The last he knew, he was being helped into a vehicle headed for Boston City Hospital. Doctors would later find claw marks on his legs, evidence of the frantic appeals of dying club patrons.

Even as people were trying to get out, others were trying to get in, screaming the names of wives, husbands, fiancées, and friends—or just wanting to help. One valiant volunteer, Gloucester sailor Stanley Viator, had been passing through the area when the fire broke out. He repeatedly dashed into the club to pull out patrons; after his fourth trip in, he did not come out. Coast Guardsman Clifford Johnson, of Missouri, had managed to escape the fire without injury. But he returned time and time again, trying to find his date. On his fourth trip out, he exited in a ball of flames. He was rushed to Boston City Hospital with second- and third-degree burns over 75 percent of his body. He was found, miraculously, to be alive on arrival, but doctors knew he wouldn't last the night. Another serviceman who rushed into the building, Navy Seaman Howard E. Sotherden, of Tiverton, Rhode Island, was in Boston on a two-day pass. Braving the choking smoke, he pulled out four people: two dead, two alive. As he pulled out one man, whose glasses still hung by one ear, he heard someone say in astonishment, "Hey, that's Marty Sheridan."

Both photos show how little actually burned in the Cocoanut Grove fire.

Huddled behind the Melody Lounge bar, Weiss slowly realized that the fire seemed to be gone. The lounge was totally quiet, although filled with smoke and a strange pungent odor. He rose and, stumbling over bodies, made it into the kitchen, where to his astonishment a group of people huddled. With them was head cashier Katherine Swett, who had been determined not to leave the money unguarded. Now certain that he knew a way out, Weiss convinced the group to follow him through the basement furnace room to a service door. But the sight of the furnace's lights and its heat spooked the group, and they ran back into the kitchen. Weiss tried to argue, but they all wanted to stay put and wait for firemen. Weiss promised to send help and dashed through the furnace room. In the searing heat, he managed to find an exit and escape into the cold air. Those who remained behind were found dead in the kitchen.

The eighteen streams of water that firefighters poured into the club appeared to be controlling the fire's terrible heat. Aerial ladders went up on Broadway and Shawmut Steets, and firemen scrambled onto the roof to vent the fire and provide access for hoses. Club staffers even helped Lieutenant Murphy find the switch that would roll back the roof. When he pulled it, nothing happened. The fuse had been removed to prevent drunken pranks. Fireman John Collins made his way into the Melody Lounge to a very pretty girl who was sitting at a table with her hand on a cocktail glass. Collins couldn't figure out why she was just sitting there; then he realized she was in a death pose.

As Engine 9 pulled up to the scene, John Crowley heard this command: Run a line from the pump to the doorway on the Shawmut Street side and get water into the building. As Crowley carried his part of the hose, he passed men and women staggering in a daze, babbling deliriously. He saw others sprawled on the sidewalk, horrifyingly still. Just concentrate on the job, his instincts told him. He and his crew dragged the hose inside the door; bodies were piled shoulder high on each side, creating an eerie passageway. As they fumbled in the dark, suddenly the lights came on, and the full impact of the catastrophe hit Crowley. He saw people still sitting at tables; they had died without even moving. He could see another pile of bodies between the Caricature Bar and the wall.

Engine 9 was ready with water, but there was no point. As of 10:45 P.M., the fire was out. The job was now rescue work—trying to find the living among the dead and moving the bodies. The company was ordered out of the dining room, and as he walked out Crowley saw one-, two-, five-, and ten-dollar bills scattered over the floor and was struck by how little money meant now. Outside he passed a squad car, its backseat filled with ladies' purses.

Through the night, the streets around the club had the surreal aura of a horror movie. Those still living were hustled into any available transportation, police cars, taxis, dump trucks, even the cars of passing motorists. Rescue workers were attempting to find any people still breathing among the limp forms on the sidewalks; they had no choice but to stack the dead bodies like cordwood. A priest moved through the crowd administering last rites. Buck Jones, recognized by his fine cowboy boots, was carried out. Elsewhere, John Walsh knelt over a crouching man who was hysterical and shaking and took him by the shoulders. The man looked up, and Walsh stared into the horror-struck face of Mickey Alpert. The man with the quick wit was speechless, fear distorting his once-dapper features.

Sheridan awoke on the dining room floor. Shaking uncontrollably but unable to move, he could hear moans and the sounds of rushing water. Then he felt someone pull him to his feet and half drag, half walk him out of the club into the bitter cold. He was helped into a cab. "Where are we going?" he gasped. "Mass General," came the reply.

Saturday night had been uneventful at Massachusetts General Hospital (MGH). Surgeons on call were listening to a football game on the radio when ambulance sirens filtered in. That was common enough. But the sirens kept coming. Grabbing their coats, the residents ran to the emergency ward. It was already filling with the wounded and the dead—men in tuxedos, women in evening dresses wearing corsages. Many of the faces were bright cherry red, a sign of carbon monoxide poisoning. Other victims had blue skin and lips, a sign of suffocation caused by oxygen being robbed from the blood. Others had froth on their lips—a sign of irritant gases. And some bore no marks at all, no burns, no discoloring; their eyes were closed as if asleep, with only dark smudges under their noses. Thomas H. Coleman, then a second-year medical student at MGH, could not believe that so many young people, seemingly unscathed, were dead. "But their flowers aren't even burned," he exclaimed. The living were gasping for breath or crying out in pain or shock from severe burns. Most were dripping wet and suffering from exposure. Dr. Francis Moore, an emergency room surgical resident, saw a naval officer, brought alive to the hospital, run frantically from room to room looking for his family. He suddenly collapsed and died, a victim of the secretions filling his lungs.

MGH, which had already developed a wartime casualty plan, put that plan to the test. Staff struggled to administer oxygen, fluids, and blood plasma; patients who were given morphine were marked with red grease-pencil Ms on their forehead. For the next forty-eight hours, nurses and doctors around the city worked without sleep.

Boston City Hospital (BCH) had an unusual number of doctors and nurses on-site that Saturday night; they had come for a holiday party. All of them left off merrymaking to pitch in when victims started to arrive. It was fortunate, because BCH received most of the victims, about 300 people. At one point, a patient was arriving every eleven seconds—a faster rate than casualties were taken to any hospital during London's worst air raids. By contrast, MGH received about 114 patients in two hours; only 39 were still alive a few hours later. Other hospitals received patients as well. In all, about 300 people were dead on arrival at hospitals, another 191 arrived, only to die later. About 440 survived. Both BCH and MGH were scrambling for beds; patients as well as bodies lined the halls. Hospital corridors soon presented as macabre a scene as that outside the nightclub itself. In critical need of blood, hospital staff lined up to donate. They also went out in the street, to pull in passersby on foot and in cars. Boston citizens responded; 1,200 citizens gave blood within days.

At MGH medical history was made—although at a horrible price. Previously, the standard treatment for burns was to scrub the skin with tannic acid to remove dead tissue and create a leathery scab to prevent infection. But Dr. Oliver Cope, then a promising researcher and later president of the American Surgical Association, believed that method was ineffectual; he believed body fluid loss and internal infection posed a greater danger than the burns themselves. He had been researching a new kind of treatment in which burns were wrapped with a gauze impregnated with a mixture of petroleum jelly and boric acid, and patients were treated intensively with intravenous fluids. Now Drs. Cope, Moore, and Bradford Cannon would apply this treatment on a large scale.

Another medical milestone was reached: medical staff also treated the burn patients with limited amounts of a brand-new substance, penicillin, to fight infection. The mold-produced agent was then considered a highly guarded secret and was used only for the military. But a thirty-two-liter supply of the drug in culture liquid form was rushed to MGH from New Jersey in a few days. Patients were also treated with sulfadiazine, from another relatively new class of agents aimed at controlling lethal blood infections. Because penicillin was administered in very low doses, its effect may not have been lifesaving. But publicity about this new "miracle" drug convinced the previously skeptical American pharmaceutical industry to start producing mass quantities of the agent. It was the beginning of the age of antibiotics.

Dr. Thomas Risley, a surgical intern at MGH, was assigned to treat the badly burned Buck Jones. As the cowboy's neck swelled, his breathing tube became blocked, and Risley and another resident spent more than two hours trying to locate his trachea for a tracheotomy. The tough hombre lingered for two days before dying from his wounds. Jones's agent, Scott Dunlap, was luckier.

Unconscious from the fumes and smoke, he awoke to find himself stacked among bodies and someone attempting to get into his pocket. "I'm alive," he managed to cry. "I'll give you three hundred dollars to get me to the hospital." A voice replied, "Where's the three hundred?" "In my wallet." Dunlap passed out again; he awoke in a hospital bed. His wallet, which had once contained $800, now held $500.

The nurses didn't think Dotty Myles would make it. She was burned over 40 percent of her body. Her angelic face was ravaged, her hair burned off to the roots, her hands nearly burned to the bones. While waiting for treatment, she begged for water; told she couldn't have any, she crawled to a water cooler and drank cup after cup. She later found out that this probably helped save her life. When she was finally put in a hospital bed, a nurse gave her a shot and mercifully she drifted into sleep.

The Nazis had him! Sheridan was sure of it. They had grabbed him, and now they were taking blood for experimentation. He couldn't see anything—his face and hands were wrapped in bandages. He could sense people bustling around him, using words like *edema* and *intravenous*. "You can't fool me with those medical terms," he whispered weakly, before sinking back into unconsciousness. Because Sheridan had given blood just weeks earlier, his blood card helped doctors begin transfusions more quickly. Hours later, when he began to gain consciousness, he recognized the voice of a friend, a doctor. He tried to spit out questions, like "Where is my wife?" "Everything is going to be all right," the doctor said nervously and excused himself. Only after a few days did Martin Sheridan's father gently tell him that Connie never made it out alive and that Buck Jones and most of the party of movie executives and their wives were dead.

Sheridan, however, was going to make it. When a doctor pulled aside the bandages on his eyes and shone a light in his eyes, he could see. When the bandages were gently unwrapped from his hand, he heard a nurse gasp, "Oh my God, we forgot to remove his ring." Sheridan's Hope High School class ring of 1932 had to be cut from his charred finger.

THE AFTERMATH

Headlines in the next day's papers screamed out the horror. "400 Dead in Hub Night Club Fire, Hundreds Hurt in Panic as Cocoanut Grove Becomes Wild Inferno," the *Globe* declared. From the *Herald*: "450 Die as Flames and Panic Trap Cocoanut Grove Crowd, Scores of Service Men Are Lost, Fire Worst in City's History, Few Victims Identified." Frantic relatives swarmed hospitals and the city's morgues, terrified of what they would find. As the bodies were slowly identified and as hospitalized victims succumbed, the death toll began to grow. District Chief William J. Mahoney's official report counted 475 deaths: "This great loss of life was, in my opinion, caused by panic. The bodies were piled four or five deep on each other and were found covered with tables and chairs, evidently upset in the mad rush for the exits and windows."

A shocked Boston could barely comprehend the immense loss in human life. Dead in the fire were the newlyweds John and Claudia O'Neill, with their best man and the maid of honor, who was the groom's sister. Grove photographer Lynn Andrews had snapped their picture during the first show, and the group decided to stick around just to pick up the photo. All four sons of 71-year-old widow Mary Fitzgerald were dead. Three young Dorchester children lost their father, mother, grandfather, grandmother and two aunts. About fifty-one servicemen and two WAVEs were dead. Grove bandleader Bernie Fazioli and singer Goody Goodelle were gone. Katherine Swett, who refused to leave her post, was found dead beside the untouched cash box. Headwaiter Frank Balzarini died in his effort to guide out patrons. Tiny Shea fought for his life. With his back and shoulders burned, he had to be immobilized facedown. He pleaded with doctors to turn him over. "Please get me off my stomach," he cried repeatedly, as nurses tried to gently explain that he needed to stay in that position. After several days of agony, Shea succumbed.

About fifteen people had survived by hiding in the walk-in refrigerator until rescued by firemen. Busboy Tony Marra found clean air by opening a freezer and sticking his head into an empty ice cream bucket. Then he heard a clang, and he raced to the walk-in refrigerator and pounded on the door, screaming: "Please let me in, I'm only fifteen years old." He heard only, "Get out of here, kid, there's no room left." Marra dashed back into the kitchen, where he spotted a window between steam pipes and, smashing the glass, wiggled to safety, maple walnut ice cream dripping from his hair.

The chorus girls escaped through the second-floor window of their dressing room, leaping into the arms of rescuers below. Others managed to get through a service door that was wrenched open. Henrietta Siegel, who danced under the name of Pepper Russell, escaped from the club unharmed. But where was her boyfriend, Al Willet, the saxophone player? Through the night, she searched for him—perhaps he was in the hospital, perhaps he

Boston Globe *headline, November 29, 1942*

had gone home. As morning approached, she forced herself to go to the Southern Mortuary. As a priest tried to comfort her there, they heard a groan from a "body." It was Al. He had been mistakenly transported among the corpses to the mortuary. He would survive.

A quirk saved Dr. Joseph Dreyfus. When the cry of fire rang out, Dreyfus stood up because, as he later explained, he was taught not to panic. He saw a sheet of flames coming across the room. Instinctively, he covered his eyes and almost instantly he passed out. He fell to the floor and wasn't discovered for hours. He later woke up in Boston City Hospital, his lungs and trachea damaged and his hands burned to the bone. The rest of his party, including his wife, were dead, some without ever having gotten up from the table. By falling to the floor so quickly Dreyfus was able to breathe cleaner, cooler air, while his party got the full brunt of the heat and the fumes.

Buck Jones's body was shipped to California. At his funeral cowboy friends and actors sang his favorite songs and Monogram Vice President Trem Carr described how Jones courageously ran back into the fire twice to rescue patrons—a bit of Hollywood hype that eased the sorrow of his grief-stricken widow and daughter. His ashes were scattered in the Pacific. Within a week of the fire, Boston movie theaters, "answering the requests of thousands," began showing Buck Jones's "last" picture, *Dawn on the Great Divide*. The last person to die from wounds suffered in the fire succumbed in May 1943. The official death toll was eventually put at 492.* Boston officials reacted with all the speed of the farmer who closes the gate after the cows have escaped. More than fifty Boston nightclubs were temporarily shut down. Fire Commissioner William Arthur Reilly launched the first of two official investigations, calling more than 200 witnesses. Included were survivors Boston Building Inspector Theodore Eldracher, decorator Reuben Bodenhorn, Boston Fire Department Wire Inspector Francis Kelly, and Mickey Alpert.

---

*Because the last death occurred after the trial, the official death toll is often listed as 491.

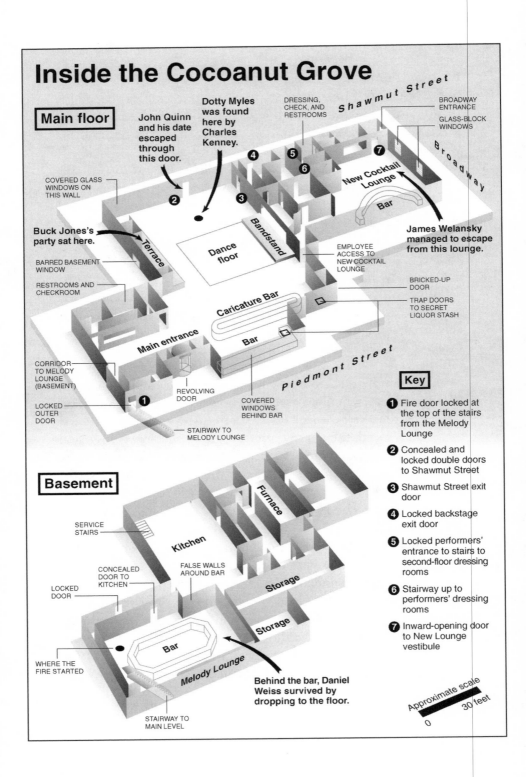

Inside the Cocoanut Grove

Main floor

John Quinn and his date escaped through this door.

Dotty Myles was found here by Charles Kenney.

DRESSING, CHECK, AND RESTROOMS

Shawmut Street

BROADWAY ENTRANCE

GLASS-BLOCK WINDOWS

Broadway

COVERED GLASS WINDOWS ON THIS WALL

2

4

5

6

3

7

New Cocktail Lounge

Bar

Terrace

Dance floor

Bandstand

EMPLOYEE ACCESS TO NEW COCKTAIL LOUNGE

James Welansky managed to escape from this lounge.

Buck Jones's party sat here.

BARRED BASEMENT WINDOW

RESTROOMS AND CHECKROOM

Caricature Bar

BRICKED-UP DOOR

TRAP DOORS TO SECRET LIQUOR STASH

Main entrance

Bar

Piedmont Street

Key

CORRIDOR TO MELODY LOUNGE (BASEMENT)

1

LOCKED OUTER DOOR

REVOLVING DOOR

COVERED WINDOWS BEHIND BAR

STAIRWAY TO MELODY LOUNGE

1 Fire door locked at the top of the stairs from the Melody Lounge

2 Concealed and locked double doors to Shawmut Street

3 Shawmut Street exit door

4 Locked backstage exit door

5 Locked performers' entrance to stairs to second-floor dressing rooms

6 Stairway up to performers' dressing rooms

7 Inward-opening door to New Lounge vestibule

Basement

SERVICE STAIRS

Furnace

Kitchen

FALSE WALLS AROUND BAR

Storage

Storage

CONCEALED DOOR TO KITCHEN

LOCKED DOOR

Bar

Melody Lounge

WHERE THE FIRE STARTED

Behind the bar, Daniel Weiss survived by dropping to the floor.

STAIRWAY TO MAIN LEVEL

Approximate scale

0   30 feet

Alpert, his hands bandaged and his face haggard, told the inquest panel that November 28 "was just another Saturday night as far as the crowd was concerned. There was a lot of gaiety. It was a football night, you know." The ebullient master of ceremonies seemed dazed; his account of his actions that night was contradictory and confusing. He said that he first realized there was trouble when singer Billy Payne said, "Hey, Mickey, it's a fight." Then "I went over to the terrace. I put up my hands and said 'Quiet.'" From there he saw flames in the lobby and then "the fire was everywhere, there was panic everywhere, screams, yells." How did he get out? There was a service door, he remembered, but it was locked. Then people pushed it open. He talked about opening another door, only to find flames behind it. He recalled breaking bars on a downstairs window and pushing people through. Then he was trying to make it up the stairs, but "I gave up. I said, 'This is it.' The next thing I know John Walsh is smacking me in the kisser."

By contrast, Joseph F. Kelley, an Arlington contractor, spoke clearly and precisely. The fire "was bluish with a yellow cast as if something were burning in suspension. It wasn't burning at any particular point. The air was full of flame yet the walls and ceiling were not then on fire." As the ball of fire passed from the foyer into the dining room, it "leveled out on the ceiling. The flames passed over my head in the passageway and went into the new lounge." Asked to describe the smoke's odor, Kelley said, "It was like nothing I have ever known. I have had a rasping feeling in my throat and chest and food hasn't tasted the same."

After Reilly's investigation concluded, he found himself on the other side of the table, answering questions from state prosecutors. Massachusetts Attorney General Robert Bushnell kicked off a criminal investigation at the direction of state Gov. Leverett Saltonstall. State Fire Marshal Stephen C. Garrity spent a year attempting to determine the cause of the fire.

Revelations soon emerged. On December 11 investigators found a bricked-up door in the wall at the end of the Caricature Bar, an exit that would have provided an outlet to those caught

on that side of the club. They also found that the club was registered in the name of Jennie Welanksy, Welansky's sister. A huge
cache of more than 4,000 cases of assorted liquor and wine was
found hidden in the ceiling and under trapdoors, minus government tax seals. The club was found to be woefully underinsured.
The original Boston Fire Department reports on the fire indicated
that the club was insured for $122,500; however, as Martin Sheri-
dan reported a year later, the six companies involved paid out
only $22,420 for damage to the club's contents. Welansky had
not bothered with getting liability insurance. The victims, who
had filed about $8 million in damages against the club, stood to
gain little. The hidden cache of liquor was auctioned, with the
intent that the $171,000 in proceeds go to victims, but the federal
government, which prosecuted Welansky for tax evasion,
declined to reduce liens against his assets. In the end, victims
received only about $150 apiece for their pain and suffering. The
hospitals bore a lot of the treatment costs, and the Red Cross
pitched in with grants and other services, but many victims faced
large medical costs.

Since a grand jury was convened, many Bostonians were convinced a massive cover-up was underway. Witnesses received
mysterious phone calls and were told to keep their mouths shut.
State Police lead arson investigator Lieutenant Philip Deady
received death threats; his son had to go to school under armed
guard for a time. Reporter Austen Lake was told by an anony-
mous caller to "go easy" on the Cocoanut Grove story, a threat
he ignored. He continued to publish a series that mercilessly
detailed the club's shady history. On December 14 his editors
printed a defiant note claiming that the threatening phone call
had been traced and "we are ready for you. The days of gang rule
in Boston are over." Even pastors railed from the pulpits for criminal indictments: "The fire has revealed a weakness in the demo-
cratic government in Massachusetts. It teaches us that
responsible citizens must be elected to public office," declared
the Reverend Samuel M. Lindsay of the Brookline Baptist
Church. John "Knocko" McCormack had another approach. A

man with some knowledge of Boston's underbelly and whose daughter died in the fire, Knocko promptly punched Mayor Tobin in the face when the mayor tried to console him during his daughter's wake.

Ten people were eventually indicted in several separate legal actions. Charged with involuntary manslaughter were the Brothers Welansky and wine steward Jacob Goldfine. Charged in separate actions, including conspiracy to violate city building codes, were Reuben Bodenhorn, Boston Fire Lieutenant Frank J. Linney, Boston Police Captain Joseph Buccigross, Boston Building Commissioner James Mooney, Building Inspector Theodore Eldracher, and contractors David Gilbert and Samuel Rudnick. Of the last eight, only Rudnick was convicted, but he never served jail time.

The Welansky manslaughter trial began on March 15, 1943. Attorney General Robert T. Bushnell brought charges in the name of 19 victims, including Adele Dreyfus. By the time the trial ended on April 10, 327 witnesses had testified and 131 exhibits were shown to the jury, including a burned palm tree, the center post of the revolving doors, and boxes of wires, carried in by State Police Lieutenant Deady. Prosecutors skirted the cause of the fire to focus on the unsafe conditions—the inadequate number of exits, the locked doors, and the unlicensed electrical work.

Welansky, defended by his law partner, Callahan, argued that patrons caused their own death by panicking instead of calmly exiting the club. In summary, Callahan told the jury, "You heard a witness say: 'It was every man for himself.' People were acting like wild animals. People were knocked down and trampled on. Many who did use their heads did get out. It was an occasion of terror, and once terror prevails, there is no telling what can happen."

In a thundering voice that reporters claimed could be heard all the way to Scollay Square, Assistant District Attorney Frederick T. Doyle responded: "To say this is an accident is ludicrous. Those responsible for that trap are sitting in the dock. . . . Someone said those who used their heads got out. James Welansky got out. Jacob Goldfine got out. But did the poor souls named here get out? It's a libel on the dead."

Moving closer to the jury box, Doyle said, "All day you have been filled up with smooth talk about panic. What did they expect the people to do? Stand up and be burned to death? They ran to one door. It was locked. They ran to another. It was locked." Doyle, who had done the bulk of the prosecution for Attorney General Bushnell, was just getting warmed up. Why did so many die? Because of "absolute greed and avarice," Doyle declared. "They were not content with an income of $1,000 a night. . . . They were not content with the average flow of trade. Instead of advertising their 'breathtaking' cocktail lounge, they should have advertised: 'Come to the Grove and abandon hope.'"

The jury was out for nearly five hours. They returned with a shocker: Jacob Goldfine: Not guilty. James Welansky: Not guilty. Barnett Welansky: Guilty.

Barnett Welansky (right), owner of the Cocoanut Grove

Of all the people involved with the Grove disaster, only the man with the master's degree in law served time. On April 14 an emotionless Welansky was sentenced to twelve to fifteen years, the "first twenty-four hours in solitary confinement and the residue at hard labor." Welansky was led away in manacles, his sentence setting a legal precedent for manslaughter cases. Welanksy didn't serve out his term. In 1946, dying from cancer, Welansky was pardoned by Maurice Tobin, who had since been elected governor. Leaving prison, he defended his release, saying, "If you were wrongfully convicted—framed—you'd feel you had a perfect right to be free." Tears came to his eyes and he added, "I only wish I had been at the fire and died with those others." He lived only another two months.

The fire made medical as well as legal history. Dr. Cope's new burn treatment had proved successful. After he and Dr. Cannon published their results in the *Annals of Surgery* in 1943, burn treatment procedures began to incorporate the "softer" petroleum jelly and boric acid approach. Other insights were gained in skin graft techniques and treatment of lung injuries. The cause of the mysterious deadly gas and strange sweet smell wasn't discovered until three years after the fire. Harvard Medical School researchers found that when the leatherette that covered the club's walls and furniture was subjected to extreme heat, it produced the lethal fume acrolein, a substance found in tear gas.

Although he escaped unscathed, Stanley Tomaszewski can be considered a victim of the fire. While others were ducking for cover, the 16-year-old busboy went to the police to tell them he lit the match that he believed started the fire. Police were impressed with the boy's honesty; he was working in the club on Friday and Saturday because the $2.47 a night plus tips that he made was needed for his ailing mother. After coming forward, he was held in police custody, primarily for his own protection. Newspaper headlines trumpeted the story about the "Busboy and the Match," and a crowd gathered outside his mother's home, screaming for his blood.

Still, witnesses did not say they saw the match ignite the tree, rather they saw sparks *after* Tomaszewski had already screwed in the light bulb. "After a careful study of all the evidence, and an analysis of all the facts presented before me, I am unable to find the conduct of this boy was the cause of the fire," Reilly declared in his final report. Tomaszewski went on with school, military duty, marriage, and a career, but he never shook the stigma of lighting the match. He once told a reporter that he "suffered the tortures of the damned." For years after the fire he and his wife received phone calls from callers who hissed, "Murderer," and vowed revenge. Tomaszewski died on October 20, 1994, at the age of 68.

But if Tomaszewski's match didn't start the fire, who or what did? Reilly and the National Fire Protection Association considered and rejected various possibilities: faulty electrical wiring, fumes from the large amount of alcohol on the premises and leaking refrigerant gases. He dismissed speculation that leftover gasoline fumes and scraps of flammable motion picture film from the building's previous incarnations as a garage and movie distribution center were to blame. Even Reilly was confounded by the bizarre nature of the fire. In his final report, which came out a year after the fire, he said, "Much of the cloth, rattan and bamboo contained in the Melody Lounge, and on the sides and lower walls of the stairway leading therefrom, was, in fact, not burned at all and the same is true of carpet on the stairway, contrary to all usual fire experience." As for the frightful speed, "the substance of the fire was a highly heated, partially burned but still burning, compressed volume of gas. By its nature this gas pressed for every available opening and I have found that this was the cause of its rapid course throughout the premises." But he found no cause. "This fire will be entered in the records of this department as being of unknown origin," he concluded.

State Fire Marshal Stephen Garrity had no answers either. "After exhaustive study and careful consideration of all the evidence, and after many personal inspections of the premises, I am unable to find precisely and exactly the immediate case of this fire," he wrote on November 8, 1943.

Speculation over the cause of the fire raged for years. Some blamed faulty wiring and combustible furnishings. The warm wall that Sheridan felt bore no effect on any of the evidence. In 1964 reporter Austen Lake published his theory that Nazi saboteurs, known to be operating on the East Coast, had set the fire. Decades later the son of firefighter Charles Kenney (also a firefighter, now retired, and also named Charles Kenney) came up with a new theory. The younger Kenney, who became a historian of the Cocoanut Grove fire, was interviewed in a flurry of stories about the fire's fiftieth anniversary. The publicity drew the attention of a retired Cambridge refrigerator repairman who, as a young man, had seen the club's refrigerator unit after it was removed from the club. Walter Hixenbaugh, now living in Florida, told Kenney he was certain that the unit was not using freon, a

Stanley Tomaszewski (left), whose lighting of a match was believed to have started the Cocoanut Grove fire, at the official inquest

nonflammable gas, as a coolant but the flammable gas methyl chloride. Freon was in short supply due to the war. He had not brought up the issue at the time because he was soon shipped overseas to fight in World War II. Oddly enough, in the Boston Fire Department's archives, Kenney had come across a December 4, 1942, letter to reporter Austen Lake from W. Irving Russell, a radio and refrigerator serviceman, who had heard Joseph Kelley's testimony. He had written to Lake to suggest that methyl chloride had been substituted for freon—but neither Lake nor Reilly had followed up on the tip. Kenney has come to believe that a refrigerator unit behind a wall in the Melody Lounge was using methyl chloride; it leaked, the gas pooled, and it was ignited by an electrical spark, thus providing the accelerant that blew through the building and created the bluish flame that Kelley described.

In 1996 Doug Beller, a National Fire Protection Association modeling specialist created a computer model of the Cocoanut Grove case that examined the methyl chloride theory, among others. Beller concluded that methyl chloride is heavier than air and thus would not explain the flames seen in the palm tree fronds. Yet he also found that when methyl chloride burns, it emits a sweet smell and releases the toxic gas phosgene, an irritant that causes lung damage consistent with the pulmonary lesions found in Cocoanut Grove victims. Still, many firefighters, including several who fought the blaze, are unconvinced, believing that the most logical explanation is still the busboy's match or perhaps bad electrical wiring coupled with the flammable furnishings. The methyl chloride theory, while intriguing, has yet to be universally accepted. Officially, the cause of the Cocoanut Grove fire remains undetermined.

There was no mystery, however, about the fire's impact on fire and building codes. Around the nation, building and safety codes were toughened. Today, revolving doors must be flanked by swinging doors or have swinging doors nearby. Exit doors must remain unlocked from the inside and must swing in the direction of exit travel. Exit signs must be clearly marked, and exits must provide a clear path to the outside. Emergency lights must have an independent power source. New Hampshire even called its 1949 updated safety codes the "Cocoanut Grove Law."

For years the shell of the once flamboyant club stood silent and boarded up. Just before it was about to be torn down, it yielded one last mystery. Someone broke into the building and emptied a safe that had been hidden in a wall under the stairs to the Melody Lounge. Welansky claimed to know nothing about it. Could it have been a cache from Solomon's days? What was in it, who opened it, and why did they wait so long to do so? These questions have never been answered. Five years before he died, Philip Deady, who had played a major role in investigating the fire, burned all his remaining notes on the case. His son Jack speculates that he kept the material as "insurance" for years. In the Cocoanut Grove case, "a lot of hands were dirty," insists Jack Deady, who has also become a historian of the fire.

Sheridan was hospitalized for two months, fighting off a series of life-threatening infections and complications. He received so many shots, he began to feel like a colander. Skin was shaved from his thighs to create grafts for his hands. Nurses had to spend two hours every day removing scabs from his face, neck, and ear. He was finally discharged in January but ordered to wear white gloves to keep infection from his still healing hands. One of his first actions was to donate blood. Another was to print up cards that he passed out liberally when asked about the white gloves: "Not that it's any of your damn business, but my hands were burned in the Cocoanut Grove and don't ask any more questions!!" He tried to work up energy for resuming his career, but he brooded about his wife, Buck Jones, and other friends who perished in the fire. He was furious that the cause of the fire was not being fully investigated. Moreover, he had to return frequently to the hospital for therapy to get his stiff hands to move again. Friends took him out to dinner, even cut his meat for him, but he was both depressed and jumpy. He left a noisy restaurant when the babble of voices reminded him of that night at the Grove. He could not shake a feeling of doom.

Sheridan was suffering from what has come to be called post-traumatic shock. The psychology of this condition and the issue of "survivor's guilt" was rigorously studied for the first time by professionals treating Cocoanut Grove survivors. The classic example involved Francis Gatturna, a 30-year-old Roslindale man whose wife had died in the fire, while he was only slightly injured. Gatturna was discharged on December 8, but on January 1 his family rushed him to the hospital, worried about his mental state. Agitated and restless, he complained of unbearable tension and trouble breathing. He repeatedly told psychiatrists Stanley Cobb and Erich Lindemann how he would have helped Grace if he had not fainted. After six days of medication and constant reassurance, he seemed to calm down. But on January 9, 1943, he hurled himself out a window and plunged several stories to his death. Gatturna's name was added to the official list of Cocoanut Grove victims.

A trip to Cuba helped cheer Sheridan up. After surviving the Grove, he decided there was really only one place for him: the war. As a correspondent for the *Boston Globe*, he was sent to the Pacific and covered action in the Philippines. Aboard the USS *Fremont* in October of 1944, 9,000 miles from home, he was approached by a young sailor. "Are you Martin Sheridan?"

"Who wants to know?" Sheridan replied.

"I'm the guy who pulled you out of the Cocoanut Grove fire," said Howard Sotherden, now an electrician's mate, first class. He explained that when he pulled out an unconscious man, he heard someone exclaim, "Why, it's Marty Sheridan." By coincidence Sheridan ended up on Sotherden's ship.

"I want a mirror. Give me a mirror. I must see," Dotty Myles begged the nurses some weeks after the fire. "If I look that bad, I have to know it. It will give me strength." Against the odds, she had survived extensive burns, severe shock, and a cardiac condition that complicated her recovery. But a Red Cross report predicted she "may never sing again and if she ever plays the piano in the future, it will not be for a long, long time." When Dotty got her mirror, what she saw did not seem too bad—her face was simply raw and red. But when scars began to appear, they turned her features into a ghastly surface of pockmarks and cracks. Webs of skin extended to her neck, and her hands were stiff and useless. She tried not to give in to despair. As the weeks went by, she endured skin graft after skin graft, lying in bed and using the rings on the bed curtain like the beads of a rosary. She often wondered who had pulled her out; sometimes she wanted to curse him, most times to thank him. Then a doctor asked her if she wanted to meet the fireman who pulled her out, and Charles Kenney walked into the room.

Charles Kenney had been missing for hours the night of the fire. His frantic wife didn't know where he might have been taken and neither did his comrades. They didn't know that he was alive—not until he walked into the station the next day to report for duty. He had regained consciousness in Boston City Hospital, groggy, coughing, but alive. Kenney struggled to his feet and made his way back to Rescue 1. Murphy took one look at him and sent him home to recuperate. He recuperated—but he never recovered.

Dotty Myles remained in the hospital six months. On her eighteenth birthday, on March 2, she even tried to sing. Though her voice was weak, she had not lost it. She was determined to sing again despite her ruined appearance. Wearing a veil and gloves, she made the rounds of Boston radio stations and scored gigs, such as a WBZ show, "Styles by Myles." She had to tolerate rude questions—such as "Did someone throw acid in your face?"—but she persevered. And then one day she was introduced to Dr. Varaztad H. Kazanjian, an Armenian immigrant who escaped the Turkish genocide. A dentist turned plastic surgeon, Kazanjian had developed new ways of treating jaw injuries. During World War I he was able to use his skill in prosthetic dentistry to reconstruct the faces of soldiers disfigured during combat. In 1941 he became the first Professor of Plastic Surgery at the Harvard Medical School. And now he was willing to help a victim of the Cocoanut Grove. After a long examination, Dr. Kazanjian's words sent a surge of hope through her:

"There's nothing to worry about so far as your face is concerned. You can be a beautiful girl again."

Dotty underwent seventeen operations to rebuild her face. By now all of Boston was rooting for her. An Air Force squadron named their B-29 after her: the *Dauntless Dotty*. Letters of support poured in. And when she finally left the hospital after the last operation, the newspapers captured her glowing smile on a smooth face.

Meanwhile, another miracle was taking place. Against all expectation, 20-year-old Clifford Johnson survived. The tenacious Coast Guardsman clung to life even as his body battled infections and internal injuries. With round-the-clock care from Boston City Hospital staff, his condition stabilized. Then began a round of intensive and painful skin grafts. Nurses spent hours rubbing cocoa butter into his skin to make it supple again. He sat up for the first time in July of 1943 and walked for the first time in September. Taken off morphine, he endured a painful withdrawal. He regained mobility, his strength, and even his boyish good looks—BCH doctors proudly showed him off at burn conferences. He returned to Missouri in 1944; on a follow-up visit to BCH, he met student nurse Marion Donovan. They were married on September 10, 1946.

But his story ends on a note of terrible coincidence. In December of 1956, while working as a park warden, his jeep rolled over, pinning him and spraying him with gasoline. Gas also hit the hot engine and ignited. Clifford Johnson burned to death.

Five days after the Cocoanut Grove fire, John Quinn was a guard of honor at Dick Vient's funeral; an hour later he was a pallbearer at Marion Luby's funeral. With his burns still healing, he shipped out a week later. One day his mail contained a pink envelope from his beloved Gerry. It began "Dear John" and indeed, she had decided she could not marry him. In true military tradition, Quinn went ashore and got blind drunk. He came home from the Pacific, a survivor of fire, war, and heartbreak. He went on to civilian life, a marriage, and children. Like many survivors, he could not speak of that night in the Grove for years. Then in 1998 his

Dotty Myles, who was badly burned in the Cocoanut Grove fire, underwent a number of surgeries and resumed her singing career.

recollection of the fire was printed in *Yankee* magazine. Days later he got a call from the man who had married Gerry. Gerry had since died, but the man wanted to thank Quinn profusely for his actions that night. Gerry's children also called and wrote to express their gratitude to him for saving their mother's life.

Other friendships endured: Sheridan continued to correspond with Sotherden and Francis Moore until their deaths. Dreyfus and Scott Dunlap kept in touch. Dancer Henrietta Siegel and her resuscitated beau Al Willet eventually married other people, but they remained friends for life.

The Cocoanut Grove fire was Charles Kenney's last fire. His lungs were permanently damaged. He also suffered from emotional scars. His son recalled how his dad once dashed out of a room in fear at the sound of an alarm—on a TV show. Kenney suffered severe coughing attacks for the rest of his life. Yet one shining light always made him feel his sacrifice was worth it. Dotty Myles remained close to the Kenney family for years. She gave Charles an autographed photo of herself, calling him "my best beau." She sent the family Christmas cards. She never forgot the sound of his voice telling her to hold on. He never failed to be filled with joy that she had survived. Kenney passed his legacy on to his children. His son Charles went on to be a firefighter, as did his grandson Tom.

Dotty Myles resumed her career; she sang in New York clubs and often returned to Boston, where, under the name of Dorothy McManus, she sang popular Irish ballads with Irish bands. One band included a young Joe Derrane, who later won national fame as an Irish accordion player. Decades later he still remembered her beautiful voice and lovely face, with only a hint of scarring on the neck and wrists. In 1960 Dotty told a New York newspaper that "people must come out of bad experiences a little better than they were. Through my own experience, I came to know myself better, and I learned about others."

Martin Sheridan,
Cocoanut Grove
survivor, today

After the fire Mickey Alpert couldn't stay in Boston. The club that for so long was associated with his name was gone. Sick at heart, he couldn't handle the memories of that horrible night and the accusing eyes of those who wondered why he survived when their loved ones didn't. He left for New York, where he married his longtime girlfriend (who had herself, ironically, been badly burned in another fire) in the home of mutual friend Milton Berle. He launched a career in a new form of entertainment, variety television, as a casting director and worked for Berle's *Texaco Star Theater*, Jackie Gleason, and Ed Sullivan. But he could never escape his memories of the fire. He never talked about that terrible night, but his daughter, Jane, said that the fire lurked in the shadows of their family life. "The fire was always part of his life. I always knew about it," she said. Mickey Alpert died in 1965, when Jane was 19. When she saw a news photograph of her father that night outside the club, the look of horror on his face haunted her for years.

George Alpert went on to become president of the New York–New Haven and Hartford Railroad and a distinguished corporate attorney. In a curious footnote, George's son, Richard, dropped out of Harvard with LSD guru Timothy Leary to become the 1960s spiritual leader Baba Ram Dass, author of *Be Here Now*.

After the war Marty Sheridan pursued a writing and public relations career, which took him from Boston to Chicago; he eventually retired to Connecticut. He married, had children, and his children had children. He never forgot the fire and he never wanted Boston to forget. On major anniversaries of the fire, even sixty years later, he badgered editors to write about it again. He still has the class ring that was cut from his finger and one of those "Don't ask questions" cards. At age 88, he now welcomes questions about the Cocoanut Grove; it's the price he pays for living while so many died.

After fighting the Cocoanut Grove fire, John Crowley wasn't sure he could report back to duty. How did the men handle this? How could he do this again? Crowley fought that inner battle and won. He stayed on the job for another thirty-one years.

Fifty years after the fire, survivors and city officials gathered for a solemn memorial ceremony. By then any hint of the club was long gone, and the configuration of the streets had dramatically changed. Now a parking garage and lot occupy what was once the Broadway Lounge and main dining area. With survivors looking on, including Tony Marra, who still recalled the taste of the maple walnut ice cream that dripped from his hair, a plaque was set in the sidewalk near the location of the revolving doors. That and a small sign on the parking garage are all that mark the spot where so many people died. From time to time, however, flowers are left near the plaque, placed there by those who never want to forget the lessons of the Cocoanut Grove.

Just after midnight, Charles Kenney—whose father fought the Cocoanut Grove fire—got a phone call from a fellow fire buff. "Charlie," the man said, his voice taut with emotion, "you're not going to believe this. I'm headed to Rhode Island to a club fire. It's another Cocoanut Grove."

An hour earlier, on February 20, 2003, the heavy-metal rock band Great White began its late-night set at The Station, a popular rock club in West Warwick, Rhode Island. Great White frequently punctuated shows with pyrotechnic displays; when they launched into the song "Desert Moon," the stage lit up with small explosions and a crowd of several hundred jammed into the one-story, wooden building roared their approval. When the fireworks set the ceiling on fire, many patrons thought the flames were part of the show. Within seconds, however, the fire began to spread, the band stopped playing, and club goers realized the building was on fire.

Just as in the Cocoanut Grove, most of the crowd tried to flee back through the door they were most familiar with—the main entrance—although some dashed to other exits and a few broke through a window. Within three minutes, the fire—apparently accelerated by flammable acoustic foam tiles—had engulfed the building. Terrified patrons became wedged in the exit doors, trapping others inside. Those lucky enough to get out tried frantically to pull others to safety.

The Station, unlike the Cocoanut Grove, burned to the ground. More than 100 people were burned to death or died later from their injuries. Great White's guitarist was among the dead. Many bodies were burned beyond recognition. Nearly 200 people were injured; some may be scarred for life. In a bizarre coincidence, a Providence TV cameraman caught the start and initial spread of the fire on camera. The club's co-owner, Jeff Derderian, worked at the same station; the cameraman was shooting a story for Derderian on nightclub safety.

The anguished questions raised in the fire's aftermath mirrored those posed sixty years earlier in Boston. Was the club packed beyond its legal limit? Had the building been properly inspected? Why did the fire spread so quickly? Grief-stricken club goers talked of seeking friends and relatives only to realize that their loved ones had never gotten out of the club. Lawmakers quickly proposed new regulations; nightclub inspections in the region were stepped up; indoor pyrotechnics were banned; lawsuits were prepared. Like the Cocoanut Grove fire, The Station fire leaves a legacy of new rules and damaged lives.

8

DEATH IN THE VENDOME

The Perils of Firefighting

After the fire is out and the company is returning to its station, the goofing begins. That's when the guys, sweaty, soot-stained, and exhausted, grab a cigarette and a cup of coffee and let themselves kick back. That's when the ribbing begins, tricks get played, and joking insults fly back and forth. It's an exercise in the rough camaraderie that gets firefighters through hard work and helps them deal with the heart-wrenching stuff they might have seen that day. Lieutenant James McCabe, who joined the department in 1958, would sometimes come to work early just for the chance to needle the guys getting off their shift. A short (five-foot-four), wiry jake, he got tagged with the nickname "Midget," but the Midget could give as good as he got. At age 43, he knew his business—he'd worked at Engine Company No. 13 in Grove Hall, a "working" station, meaning it went out to every "working" fire (a fire that requires the efforts of all responding companies). Most weeks they'd go to two to three fires a night. That's the way McCabe liked it. He wanted to be where the action was. Then in 1970 he was promoted to lieutenant and stationed at Engine 33 in Boston's Back Bay.

On June 17, 1972, McCabe arrived at the station at about 4:30 P.M., ready for his 6 P.M. shift. Only after he arrived did he learn that the Back Bay's magnificent old Vendome Hotel had caught fire. The fire was, for the most part, under control, but McCabe and other members of the company were to go down and relieve men at the scene. The hotel was only about five blocks away from the firehouse, so the guys decided to walk there.

Outside the old hotel, McCabe could see the fire engines and ladder trucks jamming Commonwealth Avenue and the alley behind the building. Smoke was still drifting from the upper floors. It was, he thought, just a normal building fire. He and firefighter Richard Magee reported to Deputy Chief John Clougherty. "Jimmy," said the chief, "Go to the top floor, see what's going on."

McCabe told his men, Ronald Holmes and Richard Powers, to stay in the lobby, but Magee said he would accompany the lieutenant, so the two men climbed to the hotel's fifth floor. Men from Ladder No. 13 and Engines Nos. 32 and 22 were there, doing overhaul, that is, they were using rakes (really poles with giant hooks) to pull down the ceiling and release the smoldering embers inside. A firefighter can tell, usually by the heat, when a fire might linger behind a wall or ceiling that has to be pulled down and "washed" with water from a hose. It can be dangerous work, because live embers and ash can rain down on the raker. Also, a ball of fire might drop down, covering the firefighter. So, for safety reasons, an officer usually stands back and watches the process.

"Hey, here comes the Midget," six-foot-four John Jameson, of Engine No. 22, sang out as McCabe stepped into the room. McCabe just chuckled. He radioed the chief to tell him there was still "real" fire in the ceiling. Clougherty radioed back to say he'd be sending up a hose line with men from Engine No. 32. McCabe knew most of the guys in the room, including Lieutenant Thomas Carroll of Engine 32. He and Carroll had been "privates" together, that is, they started as firefighters together. They were still fierce ping-pong opponents. Carroll was "raking" the fire, looking for hot spots.

"Hey, Tommy. What are you getting your hands dirty for? You're a lieutenant now," McCabe called out, grinning. Magee thought that was very funny. "I'll go over and help him," he said. He never got the chance.

There was no warning. One minute the men were standing on a solid floor, the next minute they were on an elevator going down. When the floor gave way, McCabe instantly realized what had happened—he'd fallen through floors before—and he braced himself to hit the next level. But the ride down kept going and going and going. With a terrible roar like a locomotive on a rampage, McCabe rode the collapse down five stories. In the choking rubble and thundering noise, McCabe had time to think only one thing. "This is it. I'm going to die."

The Commonwealth Mall, in Boston's historic Back Bay neighborhood, is a strip of green bordered by the divided lanes of Commonwealth Avenue. It is dotted with statues of famous—and not so famous—Bostonians, perched high on their pedestals as joggers, dog walkers, tourists, and urbanites pass by below. Near the corner of Dartmouth Street, an arc of dark granite beckons. A fire hat and coat, cast in bronze, lie across the polished stone as if someone had just tossed them there for a minute. Tourists puzzle over the nine names and the disjointed sentences chiseled into the rock: "5:20 P.M. The fire is contained. Companies start to make up, preparing to leave the building." If they look up from the curved granite and across the street, they will see an elegant nineteenth-century building, a building that became a death trap in 1972.

On Saturday, June 17, 1972—the day before Father's Day—workers were making progress on renovations to the largely empty Vendome Hotel. The building was a vestige of the Gilded Age in Boston's exclusive Back Bay. One five-story section had been completed in 1871; a second six-story section, which extended along Commonwealth Avenue, was built in 1880. Well into the twentieth century the hotel retained its aura of elegance. Its dining rooms and grand ballroom held wedding receptions and senior proms and the most glamorous social events. But after four fires in the 1960s, the elegant facade had faded and the hotel stood empty. In 1971 new owners launched major renovations designed to transform the structure from a hotel into luxury condominiums and a shopping mall. They also opened a restaurant, Cafe Vendome, on the first floor.

Sometime after 2 P.M. a worker noticed smoke coming from some upper floors in the older section of the building. When he got to the fourth floor, he heard the distinct crackle of a fire that had started to spread. The cafe was evacuated, and workers attempted to put out the fire. Meanwhile, another worker pulled Box 1571, sending out an alarm at 2:35 P.M. Engine Companies No. 33, 22, and 7 and Ladder Company Nos. 13 and 15 raced to the scene. Engine 33 firefighters tried to get to the fourth floor, where they could see the smoke, but they were blocked by plywood partitions and other construction material. Ladder 15 was positioned in the alley off Dartmouth Street, behind the hotel, and its 100-foot ladder was raised to the fourth floor. Braving the nearly unbearable heat, firefighters clambered up the steel rungs and jumped through a window into the smoking fourth floor, looking for the source of the fire. They managed to get a hose from Engine 33 up the ladder and onto the fourth floor; water started pouring into the building. Meanwhile, men from Engine 22 climbed the Dartmouth Street stairway to the upper floors, and a hose from Engine 7 was carried up an aerial ladder to another section of the fourth floor. At 2:44 P.M., a "working fire" was called in, and at 2:46 P.M. a second alarm was struck. The smoke grew heavier and thicker, and third and fourth alarms were transmitted at 3:02 and 3:06 at P.M. Eventually sixteen engine companies, five ladder companies, two aerial towers, and a rescue company joined the fight to save the Vendome.

Firefighters soon realized that the fire had been burning unobserved for some time, apparently having started in an enclosed space in a third-floor ceiling. "This building being a second-class structure, nearly 100 years old, [had] plenty for the fire to feed upon. It is considered by many to be the most hazardous type of building fire," said District Fire Chief John Vahey in his report on the fire. In the smoky chaos of the fourth floor, members of

Fire at the Vendome Hotel, which began on the upper floors

Engine 7 yanked out a section of plywood flooring and found fire below their feet. They poured water into the gap, extinguishing the fire, but raining scalding hot water onto companies below. On the fifth floor, members of Ladder 15 attempted to ventilate the fire by smashing through the roof; they had to climb out the window onto an aerial ladder when the smoke got too fierce. Still, the fire was successfully contained to the center of the building. By 5 P.M. the fire was deemed under control, but the overhaul process was just beginning; that is, firefighters still needed to find hot spots and ventilate and wash them until every bit of fire was out. News of the fire spread throughout the city, and firefighters showed up early for their shifts to relieve exhausted comrades.

One early arrival was Richard B. Magee of Engine 33. He had been a cop before he joined the department and was very pleased to exchange making arrests for fighting fires. Magee was studying for his lieutenant's test, and he had convinced his father, a twenty-five-year veteran of the fire department, to study with him, insisting it was never too late for a promotion. On his way out the door, Magee passed his son, also named Richard Magee. A slightly rebellious 17-year-old, young Richard had graduated high school and was planning to take the summer off. "Where are you going?" his son asked, noting that his father was leaving early. "There's a big fire. I'm going to let some of the guys off early. I'll see you in the morning," the older Magee said as he walked out the door. That was the last time Richard saw his father.

Another early arriver to the Vendome was Lieutenant John Hanbury of Ladder 13, who wasn't supposed to be on until 6 P.M. but who volunteered to give other members of his company a chance to rest. Engine 33 fireman Richard Powers was on his way out for a cigarette break—he always smoked Lucky Strikes—and passed Hanbury. Did he want to get a coffee? No, Hanbury decided he might as well get started. Hanbury climbed the stairs and ran into Al Feeney of Engine 22, a friend who had just helped him put in a new picture window at his Hyde Park home. Meanwhile, Powers walked down a hallway and lit his Lucky. The bad habit saved his life. He heard a noise and turned to see the building collapsing behind him. Feeney found himself in midair, crying, "Oh my God." The next thing he knew, he was buried upright in debris and it was completely dark. He could tell Fred Howell of Ladder 13 was above him, almost sitting on top of his head. Then "there were shouts of encouragement and I knew they were trying to dig me out. I don't have the slightest idea of how deeply I was buried or how long I was there. I was just scared and I prayed. That's a position I never want to be in again," he told a reporter days later.

As the dust settled, horrified firefighters could see that the southeast portion of the Vendome had collapsed. All five floors of a forty- by forty-five-foot section had fallen, bringing down a ladder and burying a ladder truck and seventeen firefighters under a two-story pile of debris. Under the direction of Deputy

Chief John O'Mara, firefighters began the desperate and danger-ous task of combing through the rubble for survivors, as the wail of ambulances rushing to the scene resounded. As if drawn by a magnet, off-duty firefighters from all over the city came to offer whatever assistance they could. Those who couldn't pitch in, owing to the small rescue area and the danger of more collapsing, waited silently in a nearby parking lot, keeping vigil for their friends and comrades. Meanwhile, the living and the dead were being pulled from the rubble. The dead were given last rites by Monsignor James Keating, the department's chaplain; the living were dispatched to hospitals. Feeney was taken out with two broken legs. Engine 21 firefighter Henry Hudson was rushed to Massachusetts General Hospital with multiple injuries. But because of the precarious piles of debris and the still-smoldering rubble, the work was agonizingly slow.

When McCabe hit the ground with a thud and the horrifying roar stopped, there was silence, the worst silence McCabe had ever experienced. Two beams from the roof had landed on either side of him, missing him but pinning him down. If he'd been any bigger, he would have been squashed. As it was, he could only move his head and one arm. But he was alive, he was pretty sure of that, and with his heart and breathing slowing down, he started to think. He tilted back his head and through the dust could see a patch of light overhead. How the hell would he get out of here? He could only wait. Then he heard noises above him—the sound of someone scrambling through the wreck-age—and the movement started to bring more rubble down. "Hey, I'm here!" he screamed, giving in to panic. And he heard the shout: "There's someone alive down there."

For forty-five minutes, McCabe stood in the rubble as men dug toward him. "Jimmy," one called, "Hang on. We've got to get a crane to get you out of there; this stuff is too heavy." At one point, he looked up to see Monsignor Keating being lowered down. Keating peered down at a face unrecognizable in the dust and dark: "Who is it?" he called, peering into the darkness. "Jimmy McCabe" came the reply. "Jimmy, let me give you last

The Vendome Hotel after the collapse

rites." "I don't need last rites," McCabe yelled back. "I need to get out of here." Suddenly he heard the shout, "Get the father out of there, it's going to collapse." Monsignor Keating disappeared and McCabe could hear the men running for cover, leaving him alone to wonder if he would make it out after all.

But McCabe didn't have to wait for a crane. Returning and digging carefully around him, rescuers managed to get him out from under the beams. Relief flowing through every vein, McCabe felt himself being lifted, hand to hand, shoulder to shoulder, up and out of the twenty-six-foot-deep hole. Outside, another firefighter leaned over his stretcher. "Jimmy, is anyone else in there?" The relief vanished as McCabe struggled to sit up. "Yes! Lower me down there and I'll show you." "No, Jimmy, you're going to the hospital." At the hospital, doctors closed the gash in his leg with thirteen stitches and washed his eyes, which were in agony from the lye used in the plaster that was released by the collapse. But he suffered no fractures, only bumps and bruises, and in a day insisted on checking out to attend the massive funeral service.

Young Richard Magee was hanging out in a park near his Brighton home, goofing with his friends, when his older sister Maureen ran up to the group. "You got to come home," she said, as Magee uttered something rude in typical teenager fashion. "No, you *have* to come home. Dad's hurt." A chill ran through the 17-year-old. He and his sister ran home, and when he got there, his grandfather was standing on the porch, his eyes moist. Then he knew, he just knew, without anyone telling him that his father was gone.

Into the night, the rescue work continued, even as hope of finding more men alive dwindled. The motor of Ladder 15, buried under the rubble, eerily continued to run until it ran out of gas. By 2:45 A.M. Sunday, the last man was accounted for. Nine firefighters were killed, the single most devastating loss to the Boston department in its history. Gone were Richard Magee and Thomas Carroll; Thomas W. Beckwith and Paul J. Murphy of Company No. 32; John Hanbury, Joseph P. Saniuk, and Charles E. Dolan of Ladder 13; and James E. Jameson and Joseph F. Boucher Jr. of Engine 22. They left eight widows and more than twenty children. At 7:45 A.M. the tapper circuit of the fire alarm office struck a special signal, 10-15, to signify the death of a firefighter.

A department chaplain brought the bad news to Jane Carroll. "He's in God's hands now." "Yes," Jane Carroll said, "but I need him, too." Eight-year-old Paul Murphy Jr. showed reporters the cardboard Father's Day card he had made but would now never give. Frances Dolan wept as she recalled that her husband, a twenty-five-year veteran, had started his shift early after learning about the fire. He told her not to call because he didn't know when he'd be home, but he promised to stop off at a bakery on the way back. Charlotte Hanbury said that when she heard the news of a fire, she had a feeling that her husband was among the victims. "He never got over the Trumbull Street fire in the South End, when a wall collapsed and five firemen were killed. He knew them all and worked with them," she told reporters. "And he never got over it. I think that from then on he had a feeling that was how it would be with him." A massive funeral was held for all nine men, attended by firefighters from all over the country.

The cause of the fire was never pinpointed, but an investigation soon targeted the cause of the collapse. About twenty years after the Vendome was built, a wall was removed to create a ballroom in the southeast section. This alteration shifted support for the upper floor from masonry walls to fifteen- and seven-inch wrought-iron beams. Significantly, a seven-inch cast-iron column served as the main support for half of the second floor and the weight-bearing wall above the second floor. On March 3, 1971, a development company purchased the aging building with the

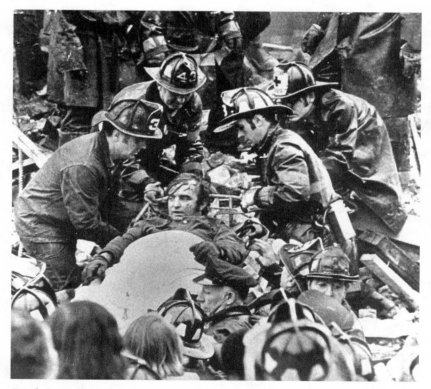

Firefighters at the Vendome attend to one of their injured colleagues.

intent of remodeling it into luxury condominiums. First, a twelve-by thirty-inch sheer metal heating and ventilation duct was installed in a basement wall; it ran about seventeen inches below the seven-inch column, according to a report by the NFPA. Later, additional renovation work was done, without proper documentation and supervision. Vahey, in his analysis of the fire, asserted that architectural plans were rushed to save money and speed the permit process. The owners, he believed, hoped to have applications filed before July 1, 1971, when a new building code would go into effect. But additional renovations, the NFPA concluded, caused excessive stresses on the seven-inch column. The dead load on this column, plus weakening created by the ventilation duct, ensured that only a small amount of weight was needed to cause the beam to fail. That weight was provided by firefighters and water from the hoses. When the beam collapsed, it brought all the floors down like a house of cards. Chief Damrell would have understood the full significance of the Vendome fire; after the great fire of 1872, he became the city's first building commissioner and spent the next twenty-six years attempting to improve the safety of construction in Boston. Barnett Welansky would have understood the urge to save money by cutting corners.

Richard Magee did not take the summer off as planned. As the oldest son, he put his personal plans on hold to help his mother and four siblings. His grandfather decided to continue studying for his lieutenant's exam in honor of his son. He took the test and won the promotion. Ten years after the Vendome fire, Richard Magee joined the fire department.

Three months after the Vendome fire, McCabe was ready to go back to work. No, he didn't want his pension. Yes, he was suffering some after-effects, but talking to the shrink provided by the department didn't help much. He just wanted to get back to work. During his first few fires, he had trouble focusing on the job; he kept imagining the walls collapsing around him. He handled his nerves by concentrating on his fellow firefighters, watching for signs of fear in their faces. If they're not running, why should I? he told himself. McCabe was eventually promoted to captain of Ladder 13 and then district chief. He continued to have coughing fits; they were a permanent souvenir of the lye burns in his lungs. And he continued to go to multiple fires during his shifts. "We always met the sun coming up on the roof of a burned-out building," he said. He retired in 1993 only because he hit the mandatory retirement age of 65. During his thirty-five years on the job, fifty-nine Boston firefighters died and he knew most of them.

Death has ever been a silent partner on fire runs in Boston. But some of the worst fire-related deaths were caused not by fire but by collapsing buildings.

In late November 1872, Chief Engineer John Damrell, confined to bed by injuries from the Great Fire of 1872, picked up a quill and dipped it into a pot of ink. *"Dear Madam,"* he wrote with a firm hand to the widow of Daniel Cochran, one of two firemen killed:

> *I embrace this, the first opportunity to address a few lines to you that I might express the grief and sorrow of my own heart. . . . Your dear husband, in order that he might save the life of another's husband, father or brother, rushed with flying footsteps to the succor of those whose voice was heard above the roaring flames and falling walls, crying for help. . . . While thus engaged in rescuing life, his spirit took its upward flight in company with those he sought to save, using the very element that was carrying destruction over our beloved city as his chariot.*

Long after the funeral speeches, the tributes, and the flowers, long after the politicians have moved on and the public has forgotten, firefighters remember and mourn their own. In Memorial Hall, a large ceremony room in the Boston Fire Department headquarters, portraits of Boston firefighters who have died on duty line all four walls. The list of the dead dates back to 1852, with tributes to John Smith, Ezra Wiley, Charles Warren, and John W. Tuttle, who all died in the 1850s. Here are Captain William Farry and Lieutenant Daniel Cochran, killed in the 1872 fire, and others who died in the 1880s, 1890s, and the twentieth century. If a photo of an early firefighter was not available, a drawing was used. If no drawing existed, just the name is there. As of this writing, 178 names and faces stare down from the walls.

There are six faces from the little-known Luongo fire of 1942. Two weeks before the Cocoanut Grove fire, on November 15, fire broke out at about 2 A.M. in the Luongo Restaurant, located on the first floor of a 100-year-old building in East Boston. A third alarm was sounded at 3:24 A.M., as companies fought to bring the blaze under control. They seemed to be gaining on it when, at 4:15 A.M. a wall of the building collapsed, trapping firefighters inside. Other firefighters and rescue workers, including the Coast Guard, rushed to the scene even as the fire regained its strength and spread to the building next door. Dozens of firefighters were pulled from the rubble, injured but alive. For eleven hours Peter McMorrow was trapped in the rubble, with fellow firefighters begging him to hang on until they could get him out. His rescue came too late. Five others also died: hosemen John F. Fole, Edward F. Macomber, Malachi F. Reddington, and Francis Degan and ladderman Daniel E. McGuire. Twenty-four-year old Degan was one of the youngest members of the department; Foley was a thirty-year veteran who planned to retire soon. More than forty firefighters were injured, including Captain John V. Stapleton, who was so bloodied by his injuries that he was thought to be dead until someone noticed he was still breathing. Stapleton was unconscious for the next week with a fractured skull and other injuries. His son remembers visiting him in the hospital and finding him barely recognizable. Stapleton recovered and returned to duty, becoming chief of department in 1950. His son, Leo, joined the department in 1951; he became commissioner and chief in 1984.

Firefighters don't think about death on a daily basis. "Depends on your point of view. A lot of us are fatalists. I know I am," former commissioner Leo Stapleton mused in an interview. He had reached the rank of captain, when a passerby spotted a fire on Trumbull Street in the South End on the night of October 1, 1964.

Fire raged in the upper floors of a vacant four-story factory as a second, third, and fourth alarm were struck; firefighters trying to get water into the building were driven back by the fierce heat. Using aerial ladders, firefighters scrambled to get water into the building. Then, without warning, part of the building collapsed, knocking men from ladders and off the balcony fire escape. As more men rushed to help the injured, another part of the building fell, burying the would-be rescuers. Five firefighters were killed in the collapse: Lieutenant John McCorkle, Lieutenant John Geswell, Francis Murphy, James Sheedy, and Robert Clougherty, the son of Chief of

The Vendome Memorial

Retired District Chief John Vahey shows a piece of paper he keeps with him at all times: a list of fellow firefighters who died in the line of duty during his years of service.

Department John Clougherty, who was also called to the fire. The father arrived only to learn that his son had been fatally injured. A freelance photographer and fire "spark," Andy Sheehan, was also killed. Five more photos soon graced Memorial Hall.

After the 1972 fire the Vendome was rebuilt. "I still get a chill when I drive by," said Richard Magee. Still, he is glad a memorial was built so close to the actual fire scene: "People in the future will know something big happened here." Every June, firefighters and other city officials gather for a brief, solemn ceremony to remember fallen firefighters. The fire has another kind of legacy: the sons of Paul J. Murphy and Charles E. Dolan also became firefighters, carrying on a family tradition.

And in the summer of 1972, nine more photos went up on the wall in Memorial Hall.

9

MR. FLARE AND THE RING OF FIRE

Arson in Boston

He called himself "Mr. Flare." He moved under cover of darkness through the streets of Boston, and where he passed, fire followed. He was no ordinary arsonist—he always seemed to stay one step ahead of authorities; indeed, he seemed to anticipate the fire department's every move. Moreover, the fires that broke out throughout the city and in neighboring counties didn't fit a pattern—or even make sense. At first the blazes of early 1982 appeared to be set to cause spectacular property damage while avoiding any possible loss of life. But firefighters were growing weary of racing to multiple serious blazes—as many as seven in a night—and hundreds of them had been injured, some badly. Arson investigators were baffled; since the breakup of a notorious Boston arson-for-profit ring in the 1970s, laws against arson had been stiffened and investigation techniques improved.

But these new arson fires defied logic. Were they the work of a serial pyromaniac? Were the fires being set to harass firefighters or to show how desperately they were needed? Could *all* the hundreds of fires set during this period be the work of one man—or two men, or a group—or copycats? Whatever the reason, by early 1982 Boston had been dubbed the "arson capital of the nation." The city would shake off that dubious distinction only after the arson ring was broken and Mr. Flare's intentions were revealed.

Arson was as old a problem in Boston as fire itself. In colonial days merchants angry over a deal sometimes sought revenge by setting fire to an adversary's home or business. As early as 1652 the Massachusetts General Court enacted a law against arson, declaring that anyone found guilty of arson could be punished by fines, whippings, banishment, or even death.

Even then some folks were simply bent on ignition. In the late 1670s many daring attempts were made by incendiaries to destroy the town. Boston fire historian Arthur Wellington Brayley suggests that, judging from their methods, a secret and determined gang of "fire bugs"' was at work.

Fire bugs set one of Boston's "great" fires, the largest fire to hit the young town up to that time. In August of 1679 the Sign of the Three Mariners tavern was mysteriously ignited, and the flames quickly spread to nearby buildings. This fire raged for thirteen hours, ultimately consuming 150 houses, stores, and warehouses and causing a loss of about £200,000. Afterward Cotton Mather lamented, "Ah Boston! Thou has seen the vanity of all worldly possessions. Fourscore of thy dwellinghouses and seventy of thy warehouses [are] in a ruinous heap." Several persons suspected of setting the fire were banished. Due to "rash and insulting speeches" he made during the fire, a Frenchman was tried for arson. He was cleared of the charges; nonetheless, he was found guilty of possession of counterfeiting tools. He stood two hours in the pillory and his ears were lopped off. He got off easy. A convicted arsonist could be whipped, branded, mutilated, or banished and assessed a fine double that of the property destroyed. An even more gruesome punishment was reserved for African slaves convicted of arson. According to Brayley,

> *A Negro woman of Mr. Lamb, of Roxbury, being indignant of some wrong done her, took revenge by setting her master's and Mr. Swan's house on fire, at midnight of July 12, 1681. The flames spread so rapidly that all the family escaped with difficulty, except one girl, who perished in the flames. The incendiary*

*was found guilty, and publicly burned to death at Boston on Sept.
22 following.*

Over the next three centuries Boston, like other major American cities, endured periods of rampant arson. Some arson fires were related to labor disputes or personal grudges. In April of 1916, groups of arsonists-for-hire were called the "Red Roosters" by Boston newspapers, reportedly because they would chalk an image of a rooster on a targeted building and use the phrase "pin a red rooster on it" as a code for igniting a blaze. Other fires were set to cover up a crime or to put a competitor out of business. Some were set by individuals who, for psychological reasons, enjoyed playing with fire; thus arson often was perceived as a crime of sexual perversity. News photographers had an old rule: Turn around and take a picture of the people *watching* the fire, because somewhere in that crowd is the person who set it.

Unlike in colonial times, the penalties for arson in the mid-twentieth century didn't involve hanging—not even the loss of an ear. Moreover, arson began to be regarded as a victimless crime—the last resort of property owners hemmed in by regulations and uncaring tenants. Boston's more notorious elements considered arson a "safe" crime; it was difficult to investigate, few prosecutions were brought, and even fewer resulted in convictions. Since fires often broke out in poorer neighborhoods, arson was deemed another hazard of urban blight, with fires set by juvenile vandals or "those kind of people"—that is, careless tenants, junkies, and hoods who had little regard for property.

So when tenants of the Fenway neighborhood of Boston began to complain about a series of suspicious fires there in the 1970s, they were greeted at first with the infuriating attitude that a poor neighborhood got what it deserved. The fires targeted apartment buildings, leaving hundreds of students, minorities, and senior citizens homeless. By 1977 there had been more than thirty fires and five deaths, mostly concentrated on Symphony Road and Westland Avenue, two densely packed blocks near Boston's famous Symphony Hall, home of the Boston Symphony Orchestra. Tired of waking up in terror at the sound of a siren in the night, a group of residents formed the Symphony Tenants Organizing Project, or STOP. The group began to complain to city officials that some kind of conspiracy was at work. Activists insisted that buildings seemed to change hands mysteriously, driving up prices and insurance coverage as "rehabilitation" was planned; then the buildings would go up in flames. One Symphony Road apartment building was sold six times in four years, the price rising from about $220,000 to $545,000. It burned down a month after the last sale. Activists contended that owners of many burned buildings had been in arrears on taxes or mortgage payments. Rather than making repairs on rundown property, they simply burned it down. Mortgage-holding banks didn't seem to mind, for loans were being paid off with insurance money.

City officials shrugged off STOP's complaints. Boston Fire Marshal Joseph Dolan insisted that he didn't see a pattern of arson of any kind, and State Police Lieutenant James DeFuria, commander of the state police arson squad, blamed the fires on grudge-settling minorities or homosexuals. "Maybe an owner does hire a torch to do a job. But you can't prove it. It's a waste of time to try," he told reporters. One of the Fenway activists, David Scondras, recalled in a television interview in 1978 that

> initially most people refused to believe that these were arson fires at all. They assumed that the kind of riff-raff that lives around here would naturally set fire to their buildings. There's some obscure notion in the minds of people that low-income people have a natural proclivity toward a variety of strange behaviors, one of them being they burn down their homes.

On September 12, 1976, fire raced through 70 Symphony Road and a 4-year-old was killed. Now both frightened and angry, STOP members began predicting publicly which buildings would burn. When some of the buildings did burn, the group's complaints finally caught the attention of Attorney General Francis X. Bellotti. But only when a Boston-based private investigation company, Boston's First Security Services, followed a paper trail that led from matchbook to checkbook did state prosecutors realize that they were looking at the country's biggest arson-for-profit conspiracy.

Founded in 1972, First Security Services started investigating suspicious fires a year later at the request of insurance companies. Larry Curran, a former military police officer and newsman, was among the First Security investigators who were learning how to track those who profit from a suspicious fire, not just how the fire itself was set. He put those skills to work in 1977, when First Security began investigating the Symphony Road fires at the request of the FAIR plan, a federal government program founded after the 1960s urban riots to help insurance companies pool resources and provide coverage for high-risk urban areas. In this case the paper trail clearly pointed to certain owners, lawyers, and contractors who benefited from the insurance payoffs. But such evidence was not enough for state prosecutors; Bellotti insisted that witnesses were needed to win a conviction. So First Security investigators paid a visit to George W. Lincoln, a low-level, wannabe wise guy who was in debt to area loan sharks and was lying low in western Massachusetts. Lincoln was known as a torch for hire; he had set fires in Boston and other cities, even burning down his Braintree home. If he cooperated, Lincoln was told, the government would grant him immunity and do right by his family. A torch but no fool, Lincoln agreed. The private investigators moved him to a hotel, and he started to spill, describing the network of people from street to boardroom who had profited from the fires. Curran was astonished at the number of individuals involved and their brazen attitude that arson was just another rehab option. Lincoln had, not surprisingly, kept a few items as protection, including canceled checks for $50 and $100 paid as bribes to none other than State Police Lieutenant DeFuria, who had investigated some of the Symphony Road fires and deemed them not suspicious. Lincoln's information also led investigators to Francis Fraine, known in South Boston as "Frankie Flame," then a wiring inspector for the city of Boston with numerous ties to Boston's underworld. Fraine had worked as a go-between in arranging the fires and also served as a lookout man. He too was persuaded to turn state's witness in return for immunity.

Attorney General Bellotti now had his witnesses. In October 1977 thirty-three men—including six lawyers, four real estate agents, four insurance adjusters, six landlords, and a city housing inspector—were arrested on charges of arson, fraud, and bribery in connection with thirty-five fires that caused $6 million in damages. Thirty-two were convicted, including DeFuria. STOP was honored nationally for helping tenants fight back, David Scondras later was elected to the Boston City Council, and Curran was invited to fire and police professional meetings to "tell the Boston story."

The breakup of the Symphony Road conspiracy didn't end arson in Boston, not even in the Fenway. But arson investigators were becoming more sophisticated in tracking both the methods and motives of fire starters—which is why investigators were so baffled by the string of suspicious fires that began breaking out in early 1982. The method, the possible motives, the pattern of destruction—none of the circumstances resembled anything they had ever investigated before. In the end, the old adage about photographing the crowd at a fire proved to be the tip that broke the case.

Boston investigators knew better, however, than to question the motives of all those who watch fires in Boston, a city with a long tradition of citizens fascinated by the drama of fighting the flames. From at least the turn of the century, Bostonians who turned out at all times of the day or night to watch a fire were dubbed "sparks"; one observer of the 1908 Chelsea fire even proudly described himself as belonging to "that great fraternity called 'sparks' and we loved to follow the engines." Some sparks were off-duty firefighters or fire volunteers, some were retired firefighters, and others were private citizens who were simply interested in fires.

Boston also has a long history of fire clubs. Veterans of the 1872 fire formed the Box 52 Association in 1873, but as the veterans aged, the association disbanded. Then, on the fortieth anniversary of the Great Fire, local business leaders revived the Box 52 Association; since then members have met regularly to discuss fire issues, raise money for fire relief efforts, and host a banquet on the anniversary of the Great Fire. The Boston Sparks Association was founded in 1938; the Tapper Club was organized in 1951. Members of such clubs represent a wide variety of interests; some sparks, for example, are primarily interested in history, others in collecting antique fire equipment and paraphernalia. By 1983 the Sparks Association had filled its headquarters at 99 West Fourth Street in South Boston with an impressive amount of fire memorabilia, including helmets, badges, books, and plaques.

But other "sparkies" chase fires with the passion of Ahab on the trail of the Great White Whale. Like nineteenth-century volunteers who would drop everything at the peal of a church bell, modern sparks rush to fires, alerted by scanners and walkie-talkies. Most try to stay out of the way of firefighters; they snap photos and discuss (or critique) the action among themselves. Some serve coffee and sandwiches to firefighters and help by carrying hoses and ladders. Some firefighters consider sparks a hindrance, others consider them a help, and still others are simply amused by it all.

One of Boston's famed sparks was Ben Ellis, owner of a fire-equipment business, who attended almost every working fire in the city. He accompanied firefighters to the Cocoanut Grove fire, where he assisted in the rescue work and gave reporters detailed descriptions of the horror. ("Suddenly, a huge sheet of flame burst through the entrance, setting fire to clothing, hair, hats, evening dresses and searing human flesh. Those flames belched 15 feet out into the street.") Ellis installed firehouse-style brass poles in his four-story office on High Street, just so he could get out of the building faster. By 1960, when Ellis was 64 years old, he estimated that he had been to more than 60,000 fires. "Fires are fascinating because they represent a struggle of the elements of man," he explained. Other famous Boston sparks have included Arthur Fiedler, the beloved music director of the Boston Pops, and David Mugar, a well-known philanthropist, who was often seen at fires with a camera in his hands. For years, sparks have relished the story of Fiedler taking television host Ed Sullivan to a spectacular warehouse fire in 1957.

The 1970s were exciting times for Boston's sparks. With large fires almost every week, sparking was a full-time hobby. Groups of buffs would gather in Roxbury at the Howard Johnson near fire headquarters on Southampton Street or at a Dorchester doughnut shop—along with night-duty news photographers and freelance shutterbugs—and wait for calls to come in. Firefighters dubbed them the "tuna fleet" because of the large radio antennas that jutted from their cars like antennas on fishing boats. United by an obsession that few outsiders understood, the men and a few women swapped information about fires they had seen and waited for the next one. It was within this circle that a 17-year-old found a refuge.

Gregg Bemis grew up in Maynard, a former mill town west of Boston. His mother, a correspondent for the local paper, began taking her son to fires when he was 4 years old. As long as he could remember, Bemis was fascinated by firefighters and firefighting. In his teens he became an auxiliary fireman for the town of Acton, and as soon as he got his driver's license, he would drive to Boston on weekends and hang out with the other sparks and race with them to fires. When he was 16, his mother died of cancer; Bemis, suffering a terrible sense of loss, set fire to the woods near the cemetery where she was buried. As he later told a reporter for *Life* magazine, he would sometimes set fires after visiting his mother's grave because it gave him a sense of release. But Bemis wanted to *be* a firefighter, so he took the civil service test, the first step to become a Boston firefighter, in 1980, when he was 19. By then he was working as a dispatcher for the Stow, Massachusetts, fire department, but compared to the almost daily fires in Boston, the work was dull. Like many before him, Bemis wanted to be where the action was.

Within the large sparking community, Bemis found both action and a sense of belonging. Sparks, who ranged in age from 20 to 70, shared his passion for fires and firefighting. The most prominent spark at the time was Elliot Belin, a man who seemed to have inherited the mantle of Ben Ellis—he got the nickname "Captain Flame" while working as an auxiliary firefighter. His unstinting zeal led some to consider him a "character and a half." Over the course of fifty years, Elliot Belin did not miss a single major blaze in Boston, and his recordkeeping rivaled that of the department itself. There was also Ray J. Norton Jr., a firefighter in his forties known to the spark crowd as "crazy Ray" and considered an oddball within the department. Although he'd been a fireman for more than ten years, he drove a car with vanity plates that spell ARSON. Robert Groblewski, another spark in his twenties, was a Boston cop, a patrolman in West Roxbury, but he really wanted to be a firefighter. So did Wayne S. Sanden, another twenty-something spark. Bemis also met Joseph M. Gorman, a rigger at the General Dynamics plant in Quincy, a quiet spark who wanted to be a state police trooper. And he met Gorman's longtime buddy, Donald F. Stackpole.

Stackpole's aura of reckless determination intrigued the young Bemis. Stackpole drove a red sedan, equipped with lights and a siren, at high speeds when chasing fires. The words FIRE DEPARTMENT EMERGENCY were printed on the tailgate, and Stackpole often wore a uniform that mimicked that of a district fire chief. He had worked as a security guard with Wayne Sanden at the state mental hospital in Mattapan and now had his own security company, Metro Security Patrol. Sanden and, later, Bemis came to work for him. Not surprisingly, Stackpole irritated most firefighters; yet he seemed to be another firefighter wannabe—albeit someone who went overboard. Unknown to Bemis, Stackpole had been convicted of receiving stolen goods, a felony, and thus could never join the fire department. Whereas Bemis, Sanden and Groblewski had all taken the civil service firefighting test (Sanden and Groblewski got perfect scores), Stackpole would forever be on the outside, nursing an inner streak of deep anger against the fire department.

Other sparks didn't like Stackpole and warned Bemis to stay away from him. Even within the admittedly obsessive sparking community, Stackpole as well as Sanden and Norton were considered extreme; in fact, the Sparks Association had repeatedly denied them membership. But Stackpole was happy to explain department procedures and the various equipment to the eager Bemis. Bemis also hung out at a fire station on Blue Hill Avenue, where he made sandwich runs, shared meals with other firefighters, tagged along on runs, and sometimes even slept there. Stephen E. McLaughlin, then a firefighter at the station, thought of Bemis as just another kid who wanted to be a firefighter. Blessed with a keen memory, Bemis soon memorized most of Boston's fire alarm box numbers; he was sure it was only a matter of time before he'd be responding to alarms wearing a BFD helmet.

Massachusetts voters had other ideas. In 1981 Proposition 2½, a statewide tax-limiting initiative, went into effect, triggering cutbacks in municipal services around the state. Boston's fire companies were reduced from 77 to 55, and about 500 firefighters were laid off. Boston police also had layoffs, and Groblewski found himself out of a job. Any chance that he, Sanden, or Bemis would be hired as firefighters faded as fire stations began to close. Safety officials complained that the cutbacks put public safety at risk; sparks began to see that firefighters were often short-staffed. Bemis was upset about a fatal fire that occurred within 100 feet of a closed fire station; he was even angrier that the rest of the city didn't seem to care. But, however justifiable his anger about Proposition 2½, there was no justifying what would happen next.

Beginning in early 1982, Boston officials started to see an increase in suspicious fires. At first the fires were simply nuisances, small dumpster fires that were easily extinguished. Then fires started breaking out in abandoned buildings, often two or three a night. On June 3, 1982, a huge fire destroyed the Spero Toy Company in South Boston, causing $13 million in damages and injuring thirty-one firefighters. On June 10 and 11, a suspicious fire broke out in a cellophane and plastics factory in Jamaica Plain; other fires were set in Dorchester and Hyde Park. On June 25 multiple-alarm fires broke out throughout the city; some of them seemed to be the work of the same firebug. "At first, it did not appear to be an organized effort," Leo Stapleton, who would later become Boston's fire commissioner, recalled in his book, *Commish*. "The city has had so many fires through the years, many of them due to arson, that it

One of the multiple suspicious fires that ripped through Boston, June 10-11, 1982

just seems that there was an unexplained increase and so what? It would die out after a while. It always did, didn't it?"

But the fires didn't die out. By mid-August, Boston had broken its record for all of 1977 with 134 multiple-alarm fires of suspicious or undetermined origin. Finger-pointing was rampant: sociologists blamed gentrification and warned darkly of corporate greed, and city officials insisted the abundance of abandoned buildings were to blame and started boarding them up. The knowledge gained from the Symphony Road arson conspiracy didn't apply; many of the buildings were not even insured, so insurance fraud couldn't explain the outbreak. Fire Commissioner George Paul told the press that the fires were the work of vandals or thrill-seeking juveniles. (He later said he couldn't be more forthcoming without jeopardizing the ongoing investigation.)

In March 1982 a variety of investigators were working together to make sense of the fires, including Stephen McLaughlin, now on the arson squad, and Rick Splaine of the BFD, Eddie Fowler of the Cambridge Fire Department, and Wayne Miller and other agents of the federal Bureau of Alcohol, Tobacco and Firearms. The men sifted through the evidence, looking for patterns of place or time, and tried to establish a profile of the arsonist (or arsonists). The fires did have a signature: the favored mode of ignition involved a cigarette stuck inside a book of matches, placed in a paper bag with paper and a plastic bag of lantern fuel. Nonchalantly smoking a cigarette, the fire starter could walk into a building, place the lit cigarette into the matches, and leave the bag against a wall. If the building was boarded up, the ingredients were put in a tire that was left leaning against a wall. Aside from the similar ignition method, the fires followed no logical pattern. A map in fire headquarters charted the fires, but just when investigators thought they had something nailed down, the pattern would shift. For example, fires seemed to be set on weekends, mostly early on Friday—then the so-called Friday Firebug started to operate during the week. Some firefighters speculated openly that the arson was the work of someone with an ax to grind against the department; others suspected disgruntled or laid-off firefighters. The fires seemed to be set in a way that would draw firefighters from other stations. Fire alarm boxes were also strangely disappearing. Usually four were stolen in a year, but in 1982 eleven went missing. Perhaps they were stolen to prevent fires from being reported quickly. Surveillance teams spread through the city on weekends from midnight to 5 A.M., but the culprits proved elusive. "We were going crazy, we were going absolutely crazy," ATF agent Miller recalled.

Almost unnoticed in the swirl of speculation and suspicion was a note, concocted out of letters cut from magazines and sent to a Boston television station in July 1982. Despite its melodramatic message and amateur execution, it held the answer to the fires. It read:

> *I'm Mr. Flare*
> *You know me as the Friday firebug.*
> *I WILL Continue till ALL*
> *deactivated police and*
> *fire equipment is brought BACK.*
> *If abandoned buildings are torn down, occupied buildings*
> *will BE targeted.*

The letter was sent by Gregg Bemis, the young man who desperately wanted to be a firefighter.

It seemed like a simple idea at first. Set a few fires and draw some attention to the need to rehire firefighters. Mr. Flare and crew started small, setting fires in dumpsters in the Back Bay, which provided a great effect when seen from Cambridge across the Charles River. But small fires don't make the evening news, so they began to get bigger and more numerous. There were rules: Fires were set only in abandoned buildings, 100 feet from an occupied building. Nobody—not even a watch-dog—was to get hurt; these arsonists were, after all, providing a public service.

It seemed almost ridiculously easy at first; with their knowl-edge of the fire department operations, the crew could easily set

Arson investigator Stephen McLaughlin with a list of all suspicious fires that occurred from June 1 to July 21, 1982

multiple fires in a night, staying one step ahead of investigators. Best of all, they could show up at the fires and mingle with the other sparks to safely watch their handiwork light up the night.

Perhaps the first fires were indeed set to draw attention to the plight of firefighters. But then it became a thrilling game of cat and mouse—the mice would show the cat that they knew more about firefighting than it did. The rules began to change. Now fires were set in empty buildings seventy-five feet, then fifty feet from occupied buildings. And though no one was killed, the fires were causing injuries and some firefighters were getting hurt badly. What started with illusions of playing Robin Hood crashed and burned on October 2, 1982, when the abandoned E Street military barracks was set on fire. Twenty-two firefighters were hurt when the roof collapsed. Manny Gregorio, an 18-year veteran, suffered two fractured vertebra; he was out of work for a year and permanently disabled. Firefighter Raymond Martin was out of work for seventeen months. Mr. Flare was concerned, but was told with a sneer that the firefighters had no business being up on the roof, that it was their own fault if they got hurt.

The game was getting out of hand. Now investigators were watching everyone who turned up at a fire, which posed a dilemma for the fire setters: If you showed up at a fire, you might be suspected, and if you didn't show up, that would be suspicious, too. So the thing to do was to drive out of town to set fires: three fires in Lawrence on September 22, four fires in Fitchburg on November 10, and others in Cambridge and Canton—that would keep them guessing. One of Stackpole's own clients even suffered a fire. Ray Norton's sparking buddies had been astonished when he suggested that they go sparking in Fitchburg, yet sure enough, that night they had plenty to see. On November 6, 1982, in a particularly brazen act, the Massachusetts Fire Academy in Stow was burned, resulting in a loss of $125,000. Personal grudges were being settled. On January 17, 1983, fire destroyed the Fourth Street headquarters of the Sparks Association, the group that refused membership to some of Mr. Flare's friends. Irreplaceable fire memorabilia were turned to ash.

Investigators finally began to suspect that certain sparks might be responsible. As McLaughlin recalled, "The light came on slow." Belin vehemently defended his friends' honor; he could not believe that any of his buddies would actually set fires. He didn't particularly care for Stackpole and he thought Ray Norton was a "total nut case," but he and Bemis had become closer. Besides, in early 1983 the fires began to trail off and gradually ceased.

By then laid-off firefighters and police, including Groblewski, were hired back under the state's Tregor Bill, which provided the city with new revenues. Sanden and Bemis, now police officers for the Boston Housing Authority, began to hope they had a chance of getting into the fire department. In 1983 Bemis took the firefighting test again.

Investigators, however, were closing in. The break in the arson gang case came during a fire at the Gerrity Lumber Company on November 21, 1982. WBZ-TV cameraman Nat Whittemore turned his camera away from the blaze to scan the gathered crowd and caught an odd group of men waving their fists in the air as if they were cheering a football team. As he aimed his camera, one of the men pulled out a gun and waved it in the air. Their behavior was so bizarre that Whittemore shared his film with authorities. They soon identified the men: Stackpole, Bemis, and Sanden. The man

Fire in the building used by the Boston Sparks Association

with the gun was a cop: Bobby Groblewski. So on November 28, ATF agents Wayne Miller and William Murphy paid a visit to Groblewski at his Weymouth home. It was one cop to another: Bobby, are you sure you don't know anything about the fires? Cool as a cucumber, Groblewski said no, he really didn't. Murphy could be cool, too. He noticed a fire alarm box in Groblewski's home. "My grandfather used to collect these; may I take a look?" he asked and took a good look at the number on the box. It matched the number of one of the stolen fire boxes. A week later, armed with a warrant, investigators returned to search Groblewski's home, charging him with receiving stolen property, in hopes of forcing him to talk about the fires. After consulting a lawyer, Groblewski clammed up.

Miller also questioned Sanden—again, one cop to another—who seemed to be trying to help but remained evasive. Sanden, however, eventually told investigators that Stackpole had stolen a car for a few parts and dumped it in the Fort Point Channel. Police divers recovered the car, but investigators still didn't have enough to tie Stackpole, Groblewski, or Sanden to the fires. Over the next year investigators tried to put the pieces together as the conspirators viewed each other with suspicion, wondering if one of them would rat them all out.

By January 13, 1984, investigators reached a pivotal moment. It was time for a full-court press on Groblewski, who had been taken off the streets and put to work as a dispatcher. Six investigators, including police officers and ATF agents Wayne Miller and James Karolides, paid a visit to Groblewski. With a theatrical flourish, Karolides started turning up the heat. "There are six people wearing badges in this room and only five deserve to," he said, looking at Groblewski. The questioning began, investigators bluffing to convince Groblewski they knew a lot more than they did. They reminded Bobby that the "first one in" would get the best deal. Miller could see that Groblewski was gradually being worn down; he sensed that the cop wanted the game to finally end, that he wanted to confess and be rid of the guilt. Groblewski dropped his head to his chest as the room fell silent.

A single word from an agent could ruin the moment. Finally Groblewski looked up. "What do you want to know?" he said simply.

For the next few hours investigators listened as Groblewski described fire after fire set by the group. By the evening's end, Miller's head was spinning and he felt sick—none of them had realized just how far the conspiracy had gone.

On February 6, 1984, Groblewski was charged with involvement in twenty-nine fires from April to November of 1982; he pleaded guilty in May and in July was sentenced to twelve years; the judge said he was puzzled that prosecutors sought a fairly light sentence. That was because Groblewski, wearing a wire, was lugging around a duffel bag full of fire slides and chatting up his buddies. He told them that yes, he was going to jail, and he just wanted to remember old times, and how about looking at some old photos? Bemis was more than happy to recall every detail. Stackpole, shrewdly, figured something was up. He wanted Groblewski to skip town, start over somewhere else. Friends from Ohio were enlisted to help spirit Groblewski safely away from investigators. But at the last minute Groblewski balked at the elaborate plans. The noose was tightening.

On July 25, 1984, Stackpole, Bemis, Norton, Gorman, Sanden, and Leonard A. Kendall of Hamilton, Ohio, an Air Force firefighter, were arrested on 83 counts of setting 163 fires that caused 282 injuries and $22 million in damages. U.S. Attorney William Weld called the conspiracy the largest arson case in state or federal history "in terms of the number of fires involved." Astounded firefighters called the men "sick puppies." Gregorio, the firefighter who had been seriously injured in the E Street Barracks fire, was floored—and then furious. "I was naturally bitter that I had been hurt," he told *Firehouse* magazine. "But I was even angrier than I would otherwise have been since public safety people were involved."

Jailed with Stackpole and Sanden, Bemis decided he had had enough when Stackpole began threatening to have Groblewski and other witnesses killed. In return for a reduced sentence,

Bemis decided to cooperate. Bemis had finally realized that "Donny was very malicious. At first I didn't see it. Then I realized he actually wanted to hurt firefighters." Based in part on his information, the charges were amended in September to 264 fires, $30 million in damages, and 360 injuries; eventually prosecutors decided to focus on 219 fires that could be absolutely traced to the gang. Authorities were able to lift a fingerprint from tape used on the Mr. Flare note; it belonged to Bemis.

The former friends faced off in court. Bemis, Sanden, and Gorman pleaded guilty and testified against Stackpole and Norton, who went to trial separately. At Stackpole's trial in late 1984, Bemis's excellent memory of the fires proved decisive. As McLaughlin said, "Every little one stuck in his memory." Bemis testified that he had initiated the fires with a two-alarm blaze in an abandoned Roxbury home but that others had rooted him on and joined in. While he provided the "modus operandi" for setting most of the fires, eventually Stackpole "more or less became the boss of the operation" and goaded and bullied the others into

Wayne Sanden (left) and Gregg Bemis conferring in Suffolk Superior Court.

continuing. When the Spero Toy Factory in South Boston burned, Stackpole bragged that "it was a fantastic fire and had overtaxed the department and really brought them to their knees," Bemis testified.

Stackpole was found guilty and sentenced to forty years. Norton went on trial in the spring of 1985. He was not charged with setting fires but with aiding and abetting the conspirators; he was sentenced to four years for his role in the conspiracy and two years for perjury. For their parts in the conspiracy, Bemis got thirty years, Sanden got twenty-six, and Gorman got five. Kendall, who was charged with participating in only one fire, was placed on probation. After he began serving his time, Gregg Bemis, now 23 years old, got wind "through the grapevine" of the results of his firefighting test: He had received a perfect score.

Maybe, as spark Elliot Belin noted sadly, the gang "did the wrong things for the right reasons," but that was cold comfort to the injured and disabled firefighters. Miller still finds it hard to fathom that grown men could have conspired for so long to set fires. None of them, perhaps, would have done it on his own, but together they played the game until events spun out of control. Even Bemis can't quite grasp why he once thought setting fires was a good idea, even if it seemed like "the lesser of two evils" at the time. As Belin said, "These guys acted like fire was a football game and they were rooting for the other team."

Twenty years later the arson fires are but a vague memory in Boston. Those convicted have served their time. As of January 2001, Elliot Belin finally stopped going to every fire (he still went to those that reached a third alarm), but he has lost none of his zest for recordkeeping; he volunteers as a running card specialist at fire department headquarters. With the continued decrease in fires and increase in Boston traffic, which has made getting quickly to a fire scene nearly impossible, the number of Boston sparks has shrunk drastically; in January of 2003 the Tapper Club disbanded.

But while fewer chase fires, the interest in firefighting history and fire memorabilia has soared. The burned-out Sparks Association met in Memorial Hall until the group found a home, happily enough, in an old fire station on 344 Congress Street. It has been turned into the Boston Fire Museum. On Saturdays from April to October, the museum is open to tourists, fire buffs, and kids intrigued by the twists and turns of Boston's trail of fire. There's an antique hand pumper, a Gamewell receiver/transmitter unit, a menu from the Cocoanut Grove, photographs of conflagrations, and Arthur Fiedler's collection of fire helmets and badges. Through the long summer afternoons, volunteer sparks answer questions from visitors and talk among themselves about Boston's continuing fight against fire.

AFTERWORD

During research for this book, while I was expounding on yet another interesting fact about Boston's fire history, a colleague shook his head. "You're becoming a spark," he insisted. I demurred. I don't have an interest in chasing fires, I said; I don't plan on getting a scanner for my car and I would never be able to rattle off facts and statistics about a fire engine as it raced by. I was, in fact, far more terrified of fires than I was before I began this research. And my fascination with fire was simply focused on the amazing history of major blazes in Boston—for example, did he know that. . . . "Yep," the colleague interrupted, with a wicked grin. "A spark."

There's no denying that the history of firefighting—like a fire itself—is filled with drama. The journey along Boston's fire trail took me into both well-known and little-known chapters of the city's history. Realizing that Boston had a "great fire" in 1872 was little preparation for its impact as recorded by the many photos and drawings of its huge swath of destruction. One afternoon, using an old map, I walked the route of the 1872 conflagration, from its origin on Summer Street to the last lines of defense near Kilby and Batterymarch Streets. I had strolled this area many times and circled the blocks in endless loops looking for parking, never realizing what history lay in their configuration. Passersby frequently approached me as I studied the old map to inquire if I needed directions; I had to explain that I was taking a walk back in time.

While much of my research came from archives and interviews, some physical action was required. When reading Chief John Damrell's testimony about his actions at the start of the 1872 fire, I puzzled over his insistence that he got to the fire within a few minutes of hearing the bells strike 52: wasn't he exaggerating? So I donned my running shoes and, starting from his home address on Temple Street on Beacon Hill, I ran his exact route to Summer Street. I made it in six minutes. Obviously, the chief was being truthful—he didn't have to dodge traffic as I did. In the small town of Dover, the local fire chief kindly allowed me to hold Damrell's helmets, donated to the department by his granddaughter. Pride and honor seem writ in the design and fading paint.

The horror of the Cocoanut Grove fire was hard to shake. Today, I cannot walk through a revolving door without checking to see if it is properly flanked by two regular swing doors. In a dim, smoky music club, I found myself looking for the exit signs and calculating the time it would take to get there . . . just in case. One of my saddest moments was looking through the *Boston Herald*'s stack of fading photos of the individual victims, among them graduation pictures, wedding portraits, and carefree snapshots. I was looking for some of the well-known victims, but I found myself lingering over every picture, marveling at the crimped hair of the young women, the boyish grins of servicemen, the elegant poise of a woman in a fur, the girl holding a cat—all people whose names do not appear in prominent accounts of the fire. The sense of loss I got from studying these photos was far greater than what I felt when I examined shots of the club's wrecked shell. What might these unknown people have accomplished? What future would they have enjoyed? Through the marvel of the Internet, I found a recording of Dorothy Myles singing "Homes Around Ballinahone," under the name of Dorothy McManus on a Rego Records recording titled *Irish American Music of the 50s*. The lilting tones and uncanny accent made it difficult to realize it was the voice of Dorothy Metzger and a woman who had endured unimaginable pain.

The largely untold story of the William F. Channing/Moses Farmer partnership was a journey into the world of Yankee ingenuity and the Athens of America. I pieced together different aspects of their lives from fire-history books, biographical encyclopedias, and the William F. Channing collection of papers at the Massachusetts Historical Society. Tracing the Channing family connections became an exercise in connecting the dots among the nineteenth century's best and brightest and showed that the hub of the solar system was a small town in disguise.

Dozens of names on the fire trail figure prominently in other volumes of Boston history: Increase and Cotton Mather, Paul Revere, Oliver Wendell Holmes, Harriet Beecher Stowe, Patrick Donohue, Elizabeth Palmer Peabody, Alexander Graham Bell, Thomas Watson, and Maurice Tobin. Likewise, certain family names ran through the list of firefighters: three generations of Kenneys have fought blazes; three generations of Stapletons have served on the fire department; two Stapletons have been chief; Gerard Crowley, the son of John F. Crowley, whose first fire was at the Cocoanut Grove, has served three decades with the department's marine unit. Several sons of firefighters who died in the collapse of the Vendome have carried on the tradition of their fathers—fully aware of the risks they are taking. The firefighting family crosses city and state lines. District Fire Chief Richard B. Magee, whose father died in the Vendome, told me how a New York firefighter, attending the dedication of the Vendome memorial, gave him an FDNY tie clip. He never got the man's name, and now he wonders whether perhaps the man was among the firefighters who died in the collapse of the World Trade Center in the September 11, 2001, attack. He now wears the pin in the unknown firefighter's honor.

I found that few fires can really be considered "minor." Whether it's a small blaze that smokes up your apartment; whether it's a larger blaze that spares your lives but takes your house, your clothes, jewelry, family photo album, television; whether you lose your pet, your car, or place of employment: any fire can be disastrous. Fire insurance can ease the financial burden; the emotional costs are not as easily remedied. No fire is "minor" if it affects your life.

Likewise, memories of fires linger in the minds of those who fought them, whether the fire took dozens of lives or a single one.

A full list of all of Boston's major fires, or even of its fatal fires, would easily fill the pages of several books. During interviews, time after time, a firefighter or buff would bring up yet another blaze that he had witnessed or that should not be forgotten.

No one was killed in the Bellflower Street fire, yet it was one of the few modern fires that could be considered a conflagration. On a warm afternoon on Friday, May 22, 1964, flames were spotted underneath a porch at 26 Bellflower Street in the Dorchester section of Boston, an area chockablock with aging triple-decker apartments. A resident called the fire department about 1:30 P.M. but almost immediately dropped the phone, saying, "I got to get out of here." The fire spread so rapidly that twenty-nine triple-deckers caught fire and a fifth alarm was sounded seven minutes after the fire was confirmed. Luckily, most residents were at work and school, but by 2 P.M. the deputy fire chief at the scene was asking for "everything available." By 4:45 P.M. the fire was under control. Nineteen apartments were gutted, leaving 300 persons homeless and causing $750,000 in damage.

Boston has had its own towering infernos. On January 2, 1986, one of Boston's tallest buildings caught on fire, endangering the lives of 1,500 office workers. The fifty-three-story Prudential Tower is a distinctive landmark on the Boston skyline. Built in 1965, the structure was not required to have a sprinkler system. About 5 P.M. on January 2, a security guard noticed that a fire alarm had been triggered on the fourteenth floor. Firefighters responding to the call could see fire in the windows of the fourteenth floor and a second alarm was struck. Other reports started flooding in, saying that smoke was not only rolling throughout the upper floors but it was banking down to the lower floors—even reaching the first floor. Meanwhile, hundreds of terrified office workers were trying to grope their way down the smoke-filled staircases. Firefighters, equipped with masks, reached the fourteenth floor through a stairwell and found the door glowing an eerie red. Because people were still trying to get down from the upper floor, the fire could not be vented from the top. So firefighters poured water on the glowing door as others climbed the stairs to help the evacuation. Sharing their oxygen, they helped the workers, many in hysterics, down the stairs to safety. About twenty-four people were injured, but no one was killed. Investigators found that the fire had started in rubbish illegally stored on the fourteenth floor, which was undergoing construction. The Prudential was retrofitted with sprinklers, and the fire led to the law that any building built higher than seventy feet in Boston had to install a sprinkler system.

Cities near Boston have also suffered major conflagrations. On June 25, 1914, fire swept through an industrial section of Salem, destroying about twenty factories and moving so fast that residents had little time to save belongings. The fire started with an explosion in one of a group of four-story wood-frame buildings and soon roared out of control. As an insurance company report later noted, "The fire department was powerless from the first." Firefighters from Peabody, Lynn, Swampscott, Boston, and other communities joined the battle, but the conflagration burned through one of the city's "best" residential sections to the waterfront and into poorer neighborhoods and the large Naumkeag Stream Cotton Company. Firefighters finally stopped the advancing flames at 2 A.M. the next day. The fire burned 253 acres—about one-third of the city—and total property loss ranged from $14 million to $16 million.

The city of Lynn, struggling with the challenge of urban renewal, was engulfed in a conflagration on November 28, 1981. The toll was extraordinary for modern times: four city blocks in the city's former shoe-manufacturing district were leveled; 17 buildings and 37 businesses providing 1,500 jobs were destroyed at a cost of $70 million. No one was killed, but 400 people were left homeless. "It looks like Berlin in 1945," Fire Chief Joseph Scanlon lamented. North Shore Community College was built in the burnt-out area.

What appeared at first to be a routine fire in Worcester just a few weeks before Christmas on December 3, 1999, became a fiery bier for firefighters on a mission of mercy. During the late afternoon, two homeless people squatting in the Worcester Cold

Storage and Warehouse Company, a former meat-packing plant, knocked over a candle during an argument and started a fire. They attempted to put out the flames themselves, but it quickly got out of control. Inexplicably, the pair ran out of the warehouse and left the area without reporting the fire. The blaze was eventually called in about 6:13 P.M. Arriving Worcester firefighters, told homeless people might be inside, donned breathing gear and plunged into the nearly windowless 110,000-square-foot building. Disoriented by smoke and the building's labyrinthine layout, two firefighters became lost and called for help as their oxygen began to run out. With the building rapidly becoming a hellish mix of heat and smoke, firefighters began a frantic attempt to rescue their two comrades. But as conditions worsened, all men were ordered out. Four of the rescuers didn't make it. After the fire was brought under control, firefighters spent the next several days combing the ruins for the bodies of the six lost firefighters; the last one was brought out more than a week later. Charges of negligent manslaughter were brought against the two homeless people; as precedent, prosecutors cited the conviction of club owner Barnett Welansky for his role in the Cocoanut Grove fire.

While Boston's fire department clings to many traditions, many aspects of firefighting have changed. Sharp-eyed readers may note the use of the non-gender-specific term firefighters in these pages—even in descriptions of historical fires, before women were on the front lines. After I was corrected several times in conversation by (male) firefighters when I used the term "fireman," I decided the generic term was more appropriate in most cases. My knowledge of other aspects of women's history makes me sure that plenty of young women would have been happy to fight alongside husbands, brothers, and fathers had the tenor of the times permitted it.

The most astonishing part of my journey along the fire trail was how close it came to my daily life. A Somerville pottery studio on Broadway where I take classes sits on the edge of what would have been the Mount Benedict convent grounds. When I wandered through the neighborhood, I wondered if I could be crossing the route that Louisa Goddard Whitney and the other students took in their mad dash from the flames; there was no way to know. When I venture out for a noontime run from my office, through the South End to the Esplanade, my route takes me only a stone's throw from the site of the Cocoanut Grove. When I buy flowers at a stand near the Old South Meeting House, I can't help but picture how it looked when flames licked at its walls. I'm aware of fire safety codes and how the 1872 fire shifted Boston's geography. I've learned how the Irish once had to battle to join the ranks of firefighters and how a rash of deliberately set fires changed preconceptions of arson. The mysteries of the Cocoanut Grove haunt me.

If this makes me a spark, so be it.

REFERENCES

PRIMARY SOURCES

Benzaquin, Paul. *Fire in Boston's Cocoanut Grove: Holocaust!* Boston: Branden Press, 1967.

Brayley, Arthur Wellington. *A Complete History of the Boston Fire Department: including the fire-alarm service and the protective department from 1630 to 1888.* Boston: J. P. Daley, 1889.

Bridenbaugh, Carl. *Cities in Revolt: Urban Life in America.* New York: Alfred A. Knopf, 1955.

Bridenbaugh, Carl. *Cities in the Wilderness: The First Century of Urban Life in America: 1625–1742.* New York: Alfred A. Knopf, 1955.

Bugbee, James M. *Fires and Fire Departments,* reprinted from *The North-American Review,* July 1873. Boston: James R. Osgood and Company, 1873.

Cannon, Donald J. *Heritage of Flames: The Illustrated History of Early American Firefighting.* Garden City, N.Y.: Artisan Graphics Corp., 1977.

Conway, Fred. *FireFighting Lore: Strange but True Stories from Firefighting History.* (Foreword by Paul Ditzel.) New Albany, Ind.: Fire Buff House Publishers, 1993.

Conwell, Russell Herman. *History of the Great Fire in Boston, November 9 and 10, 1872.* Boston: B. B. Russell; Philadelphia: Quaker-City Publishing House, 1873.

Dana, David D. *The Fireman: Fire Departments of the United States, with a Full Account of All Large Fires, Statistics of Losses and Expenses.* Boston: James French and Company, 1858.

Ditzel, Paul C. *Fire Engine, Fire Fighters: The Men, Equipment and Machines from Colonial Days to the Present.* New York: Crown Publishers, 1976.

Ditzel, Paul C. *The Fascinating Story Behind the Red Box on the Corner.* New Albany, Ind.: Fire Buff House Publishers,1990.

Greenberg, Amy S. *Cause for Alarm: The Volunteer Fire Department in the Nineteenth-Century City.* Princeton, N.J.: Princeton University Press, 1998.

Haywood, Charles F. *General Alarm: A Dramatic Account of Fires and Fire-Fighting in America,* New York: Dodd, Mead and Company, 1967.

Hazen, Margaret Hindle, and Robert M. Hazen. *Keepers of the Flame: The Role of Fire in American Culture 1775-1925.* Princeton, N.J.: Princeton University Press, 1992.

Holzman, Robert S. *The Romance of Firefighting.* New York: Harper & Brothers, 1956.

Keyes, Edward. *Cocoanut Grove.* New York: Atheneum, 1984.

Lord, Robert Howard, John E. Sexton, and Edward T. Harrington. *History of the Archdiocese of Boston in the Various Stages of its Development, 1604 to 1943.* New York: Sheed and Ward, 1944.

Lyons, Paul Robert. *Fire in America.* Boston: National Fire Protection Association, 1976.

Morris, John V. *Fires and Firefighters.* Boston: Little, Brown and Company, 1953.

O'Connor, Thomas. *Eminent Bostonians.* Cambridge, Mass.: Harvard University Press, 2002.

Pratt, Walter Merriam. *The Burning of Chelsea.* Boston: Sampson Publishing Co., 1908.

Schultz, Nancy Lusignan. *Fire and Roses: The Burning of the Charlestown Convent, 1834.* Boston: Northeastern University Press, 2000.

Seaburg, Carl. *Boston Observed.* Boston: Beacon Press, 1971.

Smith, Dennis. *History of Firefighting in America: 300 Years of Courage.* New York: Dial Press, 1978.

Stapleton, Leo. *Commish.* Boston: DMC Associates, 1990.

Stapleton, Leo. *Thirty Years on the Line.* Boston; Quinlan, 1983.

Tager, Jack. *Boston Riots: Three Centuries of Social Violence.* Boston: Northeastern University Press, 2000.

Tufts, Edward R. *Hunneman's Amazing Fire Engines.* New Albany Ind.: Fire Buff Publishers, 1995.

Werner, William. *History of the Boston Fire Department and Boston Fire Alarm System, January 1, 1859 through December 31, 1973.* Boston: Boston Sparks Association, 1974, 1975.

Werner, William. *The Evolution of the Boston Fire Department, 1678–1977.* Dallas: Taylor Publishing, 1977.

Whitney, Louisa. *The Burning of the Convent: A narrative of the destruction by a mob of the Ursuline school on Mount Benedict, Charlestown, as remembered by one of the pupils.* Boston: Osgood, 1877.

Zurier, Rebecca. *The Firehouse: An Architectural and Social History.* New York: Artabras Publishers, 1982.

OTHER SOURCES

The Drukman Collection, Cocoanut Grove, vertical file, Bostonian Society.

Putnam Diaries, William F. Channing collection, Massachusetts Historical Society.

REFERENCE NOTES

CHAPTER 1 BUILT TO BURN

1 *Mary Morse:* Described by Brayley, p. 29, and Josiah Henry Benton, *The Story of the Old Boston Town House, 1658-1711* (Boston: private printing, 1908), quoting Judge Samuel Sewall's diary that fire broke out "by reason of the Drunkenness of _____Moss," p. 207. Spelling in colonial times was variable and creative.

1 Background on early Boston: Benton, *Story of the Old Boston Town House;* Bridenbaugh, *Cities in Revolt;* Bridenbaugh, *Cities in the Wilderness;* Seaburg, *Boston Observed;* John Jenning, *Boston: Cradle of Liberty* (Garden City, N.Y.: Doubleday, 1947).

1 *"as already happened":* Jenning, p. 118.

2 *"The Flames took Hold":* Benton, p. 206.

3 *"Many poor men":* Ibid., p. 211.

3 *"Desolating fires":* Ibid., pp. 207, 209; Ditzel, *Fire Engine,* p. 23.

4 *"Boston was built to burn":* Ditzel, *Fire Engine,* p. 8.

4 *"About noon":* Winthrop diary, cited by Brayley, pp. 3–4.

4 *"we have ordered":* Dudley diary, cited by Ibid., p. 4.

5 *"Here is good living"*: William Cronon, *Changes in the Land: Indians, Colonists, and the Ecology of New England* (New York: Hill and Wang, 1983), p. 25.

5 *"My Ink in my very pen"*: cited by Hazen, p. 106.

5 *"It was a wonderful favor of God"*: This often-cited quote is of unclear origin. Brayley attributes it to Governor Winthrop, who had died by that date; Cannon attributes it to succeeding Governor Endicott; and Seaburg attributes it to Winthrop's son, who became governor of Connecticut.

5 Boston's early fire regulations: Brayley, p. 6; Benton, pp. 205–206; Lyons, pp. 3–6; Cannon, pp.18–19, 24–25.

5 *"If fires could be legislated"*: Ditzel, *Fire Engine,* p. 21.

6 No smoking outside: Ibid, p. 18.

6 Construction guidelines: Brayley, p. 6; Benton, p. 205.

6 *"Oh! Lord God"* and speculation on foreknowledge of fire: Cannon, pp. 54–55.

6 *more fires than all other American cities combined:* Bridenbaugh, *Cities in Revolt,* p. 100.

6 *"At the rate Boston burned"*: Ditzel, *Fire Engine,* p. 15.

7 Early firefighting equipment and organization: Cannon; Ditzel, *Fire Engine;* Smith.

8 *twenty years before Paris*: Bridenbaugh, *Cities in the Wilderness,* p. 209.

8 *"immediately be clipt into Gaol"*: Bridenbaugh, *Cities in the Wilderness,* p. 211.

8 Background of early fire societies: Brayley, pp. 37–38; Zurich, p. 39; Bridenbaugh, *Cities in Revolt,* p. 294.

9 Benjamin Franklin background: Bridenbaugh, *Cities in the Wilderness,* p. 369; Cannon, pp. 91–94.

9 Peter Stuyesant background: Ditzel, *Fire Engine,* p. 20.

10 *"beheld a blaze"*: cited by Seaburg, p. 146.

10 *"The distressed inhabitants"*: Brayley, p. 60.

10 *"I can say without exaggerations"*: Seaburg, p. 127.

10 Losses from the 1760 fire: Smith, p. 20.

11 *"Once renown'd beloved City"*: cited by Cannon, p. 59.

12 *"stimulating genius"*: Massachusetts Charitable Fire Society: *Two Hundredth Anniversary,* p. 8.

12 *"seemed doomed to be destroyed"*: Dana, p. 26.

12 *"an affront to the mechanics of Boston"*: Howard Mumford Jones and Bessie Zaban Jones, *The Many Voices of Boston: A Historical Anthology: 1630-1975* (Boston: Little Brown and Company, 1975), pp. 158–161, quoting Edmund Quincy's *The Life of Josiah Quincy.*

CHAPTER 2 FIRES OF WRATH

13 The primary eyewitness source for the convent fire was Louisa Goddard Whitney, who recalled her experience in detail in *The Burning of the Convent.* Other eyewitness accounts are from Lucy Thaxter White in her letter to G. T. Curtis, *Boston Evening Transcript,* February 4, 1843; from *Life of Mother St. Augustine O'Keefe, Superioress of Ursuline Convent by an Ursuline Nun* (New Orleans: s.n., 1888); and from unpublished material translated from *Les Ursulines des Trois Rivieres,* v. 2.219–231, stack 271.9, u82, vertical file, Somerville (Mass.) Public Library. Background information came from Schultz, *Fire and Roses,* and additional articles by Schultz, including "Mary Anne Moffatt's Elusive Life," *Boston College Magazine,* Fall 2000; "The Ursuline Convent Riot;

Charlestown, Massachusetts, 1835," *Sextant*, Vol. IV, November 2, 1993; and "The Ursuline Convent Riot: Charlestown Massachusetts," a publication for the Somerville Museum.

14 Anti-Irish sentiment: Tager, pp. 120–142; Jules Archer, *Rage in the Schools* (New York: Branden Press, 1994), pp. 27–32.

15 Background on Irish population in Boston: Thomas O'Connor, *The Boston Irish: A Political History* (Boston: Northeastern Unversity Press, 1995), p.37.

16 *"a woman of masculine appearance and character"*: George Hill Evans, Somerville Public Librarian, "The Burning of the Mount Benedict Ursuline Community House," unpublished historical monograph, 1934, p. 8.

17 *Reed, an obviously troubled woman:* See Rebecca Reed and Maria Monk, *Veil of Fear: Nineteenth-Century Convent Tales,* introduction by Nancy Lusignan Schultz (West Lafayette, Ind.: NotaBell Books, 1999).

18 *Placards began to appear:* Evans, p. 10.

18 *"The principles of this corrupt church"*: Lyman Beecher, cited in Schultz, *Fire and Roses,* p. 166.

19 *"Ironically"*: Ibid., p. 19.

19 *"a dense black mass"*: Lucy Thaxter letter.

20 *"Disperse immediately"*: Whitney, p. 113. Accounts differ as to the exact wording of the Mother Superior's challenge.

20 *determined to "clean the establishment out"*: Buzzell's comments are from trial transcripts quoted by Schultz and Jeanne Hamilton, "Nunnery as Menace: The Burning of the Charlestown Convent, 1834," *U.S. Catholic Historian,* Vol. 14, Winter 1966.

21 *"They will certainly help us"*: Whitney, p. 121.

21 *"in a spirit of deviltry"*: Schultz, *Fire and Roses,* p. 181.

21 *"They desecrated the mortuary chapel"*: Les Ursulines des Trois Rivieres.

22 *"Could such things be?"*: Lucy Thaxter letter.

23 *"Ladies, if any among you"*: Life of Mother St. Augustine, pp. 4–6, also described in *Les Ursulines des Trois Rivieres.*

23 *"Enough engines responded"*: Schultz, *Fire and Roses,* p. 174.

23 *"The testimony against me"*: Hamilton, "Nunnery as Menace."

24 *Another rioter, Benjamin Wilbur:* Lord, Sexton, and Harrington, *History of the Archdiocese of Boston,* pp. 234–235.

24 Details of the Broad Street Riot from Brayley, pp. 197–199; Tager, pp. 120–114; Lord, Sexton, and Harrington, pp. 205–239; Dana, pp. 333–336. Newspaper accounts of the Broad Street Riot include *Boston Daily Advocate,* June 12, 1837, *Independent Chronicle & Boston Patriot,* June 17, 1837, *Boston Daily Atlas,* June 17, 1837, and *Saturday Evening Gazette,* June 17, 1837, excerpts collected by Bostonian Society in preparation for the exhibit "Burning Issues: A History of Boston Through Fire," fall 2001–spring 2002, Old State House Museum.

25 *"who lingered longer"*: Lord, Sexton, and Harrington, p. 244.

26 *"A gang of stout boys and loafers"*: Brayley, p. 198.

26 *"actuated by the vindictive and destroying spirits"*: Independent Chronicle & Boston Patriot, June 17, 1837, excerpted by Bostonian Society.

26 *"We cannot but allow"*: Saturday Evening Gazette, June 17, 1837, excerpted by Bostonian Society.

26 *"The plebeian firemen"*: Tager, p. 121.

CHAPTER 3 BUILT IN BOSTON

30 Background on the idealized image of firefighters from Holzman, *The Romance of Firefighting;* Hazen and Hazen, *Keepers of the Flame;* Ditzel, *Fire Engine;* Greenberg, *Cause for Alarm.*

31 *Currier and Ives:* Greenberg, pp. 18–21.

32 *"Volunteers in the service of beneficiaries":* Townsend, Massachusetts Charitable Fire Society: *Two Hundredth Anniversary* (Boston: Massachusetts Charitable Fire Society, 1992).

32 *"With a fireman's love of 'the machine'":* Greenberg, p. 73.

32 *ardor that . . . bordered on mania:* Ibid., pp. 70–75.

32 *"We Come, We Conquer":* slogan painted on "Granite" Hunneman engine on display at National Fire Protection Association, Quincy, Massachusetts; other slogans provided by Earl Doliber.

32 *"the Hunneman was king":* Edward R. Tufts, *Hunneman's Amazing Fire Engines,* back cover. Tufts is also the source of other details of Hunneman's life.

34 *first fire "injine":* Brayley, p. 7; Ditzel, *Fire Engine,* p. 21.

35 *Possession of the finest available fire defense:* Bridenbaugh, *Cities in the Wilderness,* p. 209.

35 *"These engines are . . . warranted":* Tufts, p. 13.

36 *"A visit to these works is a rich treat":* Ibid., p. 18.

36 Prices of Hunneman engines: author's interview with J. Richard Hunneman, great-great-great grandson of founder, October 2002.

37 *offered a £5 bonus:* Holzman, *Romance of Firefighting,* p. 61.

37 *"Each man strove mightily":* Ibid., p. 3.

39 *In 1831 Boston wanted a new engine:* Tufts, p. 16.

39 *Huge numbers of spectators:* illustration, Holzman, p. 65.

39 *"The great body":* Dana, p. 12.

40 *"We are good neighbors":* Ibid., pp. 17–18.

41 *They filled their rig with water:* Ditzel, *Fire Engine,* p. 62.

41 *"I esteem the fire department":* In the Matter of the Application of Wm. F. Channing and Moses G. Farmer, for the Extension of Letters Patent, issued to Wm. F. Channing, as assignee of himself and Moses G. Farmer, May 19, 1857, 17355 (Boston: A. Mudge, 1871), Massachusetts Historical Society, p. 27.

41 Background on initial opposition to the introduction of steam engines: Ditzel, *Fire Engine,* pp. 102–113; Greenberg, pp. 139–143.

43 *"The sight of a steam fire engine":* Werner, *Evolution of the Boston Fire Department,* p. 34.

45 *"famous Boston fire department horses":* undated clipping from *The Firemen's Standard,* Drukman collection, Bostonian Society.

45 *"There was no greater opponent":* Tufts, p. 67.

47 Background on modern musters: interviews with Earl Doliber and newsletters of the OKO's Veteran Firemen's Association of Marblehead and J. Richard Hunneman; also the New England States Veteran Fireman's League of Newburyport, Massachusetts, and Handtub Junction USA (www.handtubs.com), which is "dedicated to those gallant machines of old."

CHAPTER 4 STRIKE THE ALARM

50 *"Almost all cities":* "Morse's Telegraph for Fire Alarms," *Boston Advertiser,* June 3, 1845.

50   Details on William F. Channing's family and life: obituary, *Boston Globe,* March 20, 1901; obituary, *Boston Evening Transcript,* March 20, 1901; William Ellery Channing, *Memoir, with extracts from his correspondence and manuscripts* (Boston: Wm. Crosby and H. P. Nichols, 1848); Arthur W. Brown, *William Ellery Channing* (New York: Twayne Publishers, 1962); Amalie M. Kass, *Midwifery and Medicine in Boston: Walter Channing M.D. 1786–1876* (Boston: Northeastern University Press, 2002); Thomas H. O'Connor, *Eminent Bostonians* (Cambridge: Harvard Community Press, 2002); *Dictionary of American Biography Base Set, American Council of Learned Societies, 1828–1936,* reproduced in *Biography Resource Center* (Farmington Hill, Mich.: The Gale Group, 2002); *The National Cyclopedia of American Biography* (New York: J. T. White, 1898–1984).

51   *William "did not incline to read":* Elizabeth Palmer Peabody, *Reminiscences of Rev. Wm. Ellery Channing, DD* (Boston: Roberts Brothers, 1880), pp. 259–264.

52   *"Electricity is entering":* William F. Channing, *The Medical Application of Electricity,* 5th ed. (Boston: Thomas Hall, 1859), p. 236.

52   *"most remarkable invention":* Daniel J. Czitrom, *Media and the American Mind: From Morse to McLuhan* (Chapel Hill: University of North Carolina Press, 1982).

52   *"a perpetually higher co-operation":* William F. Channing, "The American Fire-Alarm Telegraph: A Lecture," *Ninth Annual Report of the Smithsonian,* March 1855, p. 6.

53   *"The Electric Telegraph in its common use":* William F. Channing, "On the Municipal Electric Telegraph especially in its application to Fire Alarms," extracted from *American Journal of Science* (New Haven: B. L. Hanlen, 1852), p. 4.

53   Details on Moses Gerrish Farmer's life: obituary, *Boston Transcript,* May 25, 1898; *Dictionary of American Biography Base Set; American National Biography,* general eds. John A. Garraty and Mark C. Carnes (New York: Oxford University Press, 1999); *National Cyclopedia of American Biography; Encyclopedia of World Biography,* 2nd ed.

54   *Their salutations changed:* letters in William F. Channing Papers, 1851–1898, Massachusetts Historical Society.

54   *extremely detailed presentation:* "Communication from Dr. Wm. F. Channing respecting a System of Fire Alarms, read and referred to Joint Standing Committee on Public Building . . . in Common Council," March 27, 1851.

55   *"like spider webs":* "The Municipal Telegraph," *The Commonwealth,* December 30, 1851.

55   *"The act is so simple":* Channing, "On the Municipa Electric Telegraph," p. 14.

56   *The log reads:* Werner, *History of the Boston Fire Department,* p. 183.

56   *"Its utility was disbelieved":* In the Matter of the Application," p. 27

57   *Channing "made every effort":* Letter Patent application, p. 20.

58   *"it is a higher system of municipal organization":* Channing, "The American Fire-Alarm Telegraph," p. 19.

58   Details of Gamewell's life: obituary, *Hackensack Republican,* July 23, 1896; Paul Ditzel, *The Fascinating Story Behind the Red Box on the Corner* (New Albany, Ind.: Fire Buff House Publishers, 1990).

59   Background on improvements in Boston's fire alarm system: Werner, *History of the Boston Fire Department,* pp. 177–195.

63   *"I need all"* and *"I will repeat":* letters between Farmer and Channing, Channing Papers.

63   *Watson described how Farmer dropped by:* Thomas A. Watson, *Exploring Life: The Autobiography of Thomas A. Watson* (New York: D. Appleton, 1926), p. 97.

64 *"He was deserving of more honor":* Dictionary of American Biography, p. 280.
64 *"invented not wisely, but too well":* Encyclopedia of World Biography [electronic version], 2nd ed. (Detroit: Gale Research, 1998).
64 *"immanence of God":* note dated December 20, 1900, in Channing Papers.

CHAPTER 5 THE GREAT FIRE OF 1872

Primary sources for accounts of the 1872 fire include Brayley, *Complete History of the Boston Fire Department;* W. F. Chandler and Co., *Chandler & Co.'s Full Account of the Great Fire in Boston and the Ruins* (Boston: W. H. Chandler & Co. , 1872); *Report of the Commissioners Appointed to Investigate the Cause and Management of the Great Boston Fire* (Boston: Rockwell & Churchill, 1873); "Carleton" (Charles Carleton Coffin), *The Story of the Great Fire, Boston, November 9–10, 1872* (Boston: Shepard and Gill, 1972); Conwell, *History of the Great Fire in Boston;* Charles Stanhope Damrell, *A Half Century of Boston's Buildings* (Boston: L.P. Hager, 1895); Haywood, *General Alarm;* Harold Murdoch, *Gentleman in Boston: Letters Written by a Gentleman in Boston to His Friends in Paris: Describing the Great Fire* (Boston: Houghton Mifflin, 1909); Michael J. Novack, *Photography and the Great Fire of November 1872,* monograph, 3rd rev. ed. (1992); Anthony Mitchell Sammarco, *The Great Boston Fire of 1872* (Dover, N.H.: Arcadia Publishing, 1997); Robert Taylor, *The Great Boston Fire, 1872: A Disaster with a Villain: Old-Style Politics* (Boston: Boston Sunday Globe, 1972); John P. Vahey, monograms on "The Epizootic Fire" and "The Great Boston Fire," Boston Fire Department files; William Werner, "Boston's Fire Alarm Box 52 and Related Matter," document dated February 20, 1972, Boston Fire Department files.

69 *"For God's sake, hold the corner":* Damrell's testimony, *Report of the Commissioners,* p. 86. Direct quotes from Damrell are from his testimony.
69 *Mrs. O'Leary's cow:* The Boston fire's contribution to song has been the virtually unknown tune by C. A. White, "Homeless Tonight: Boston in Ashes" published by White, Smith & Perry, 1872.
69 Background on Damrell: obituary, *Boston Transcript,* November 4, 1905; Vahey, "The Epizootic Fire"; interview with John P. Vahey, June 2002..
70 *"Do not try to magnify":* Damrell's testimony, *Report of the Commissioners,* p. 105.
70 *"The hard-headed Yankee gentlemen":* Harold Murdoch, "The Great Boston Fire and Some Contributing Causes," paper presented to the Bostonian Society, November 19, 1912, published January 2, 1913, p. 53, in the collection of Boston Historical Society.
72 *"what proved to be a serious blunder":* Harold Murdoch, "How the Fire Became a Conflagration," *Current Affairs,* October 30, 1922, p. 14.
72 *On November 9, many Bostonians strolled:* Descriptions of life in Boston from Murdoch, *Gentleman in Boston,* pp. 20–29; Frank Wright Pratt, *Virgil Drops His Cane: A Story of Boston in the Seventies and the Great Fire* (Boston: Christopher Publishing House, 1938); Jane Holtz Kay, *Lost Boston* (Boston: Houghton Mifflin, 1980).
73 *"Some spark snapping outward":* Conwell, p. 54.
75 *But it was too late:* Damrell's testimony, *Report of the Commissioners,* p. 124.
77 *"a malignant recklessness":* Harper's Weekly, November 30, 1872.
77 *At about 8 P.M., Sarah Putnam:* This and subsequent descriptions are from Sarah Putnam's unpublished diary, entries for November 1872, 1860–1912, Vol. 11, collection of the Massachusetts Historical Society.

78 *When Gilman Joslin's lecture was interrupted:* This and subsequent quotes are from Joslin's letter to his brother Will, November 13, 1872, vertical file, Bostonian Society.

78 *"war dance of the fire-fiends,"* Conwell, p. 57. Other lurid descriptions include "Niagara of destruction," Brayley, p. 275; "lurid light of a burning cauldron," *Boston Journal,* November 14, 1872 ; "weird, strange dance of the devourer," *Chandler & Co.'s Full Account.*

80 *"perfect tinder boxes":* Bird's testimony, *Report of the Commissioners,* p. 522.

80 *"I don't understand it today":* Damrell's testimony, *Report of the Commissioners,* p. 114.

82 *"A building all dark":* Murdoch, *Gentleman in Boston,* p. 75.

82 *"I saw . . . the fire":* letter from Oliver Wendell Holmes to John Lothrop Motley, November 16, 1872, cited in John T. Morse, *Life and Letters of Oliver Wendell Holmes* (Boston: Houghton Mifflin, 1896), vii, p. 197.

82 footnote on Phillips Brooks: Susan Wilson, *Boston: Sites and Insights* (Boston: Beacon Press, 1994), p. 81.

84 *"I'm sorry to say they were from Cambridge":* Eliot's testimony, *Report of the Commissioners,* p. 377.

84 *"Never did a body of men work":* Damrell's testimony, *Report of the Commissioners,* p. 111.

84 *"What man could do":* Conwell, p. 59.

85 *"The fire is destined to sweep the whole city":* Burt's testimony, *Report of the Commissioners,* p. 387. This and subsequent quotes are from Burt's testimony.

86 *General Burt . . . talked a great deal":* Gaston's testimony, *Report of the Commissioners,* p. 562

87 *"I was running for my own preservation":* George O. Carpenter's testimony, *Report of the Commissioners,* p. 213.

88 Description of the confrontation between Woolley and Benham: Woolley's testimony, *Report of the Commissioners,* pp. 161–162. Benham's testimony is rather different; see *Report of the Commissioners,* pp. 639–641.

91 *"She burnt majestically":* Brooks, cited by Murdoch, *Gentleman in Boston,* p. 155.

91 *The fire devastated Boston's leather goods . . . industry:* F. E. Frothingham, *The Boston Fire, Nov. 8th and 10th, 1872: Its History Together with the Losses in Detail of Both Real and Personal Estate* (Boston, Lee & Shepard; New York, Lee, Shepard & Dillingham, 1873); Augustus Thorndike Perkins, *Losses to Literature and Art by the Great Fire in Boston* (Boston: David Clapp and Son, 1873).

91 *"The fire leaps":* Coffin, p. 16

92 *"Dear old church":* Boston Journal, November 14, 1872; Murdoch, *Gentleman in Boston,* p. 91; *Chandler & Co.'s Full Account of the Great Fire.*

93 *"The gas has been shut off":* "Letter of Eye-Witness Describes Boston Fire, 64th Anniversary of Which Occurs Next Monday," unidentified newspaper clipping about letter sent by Grace Revere to Lizzie Spooner, November 7, 1936, vertical file, Bostonian Society.

94 *"It will be remembered":* Coffin, p. 29.

94 *"through the ghastly piles":* Conwell, p. 104.

94 *Sixteen other people . . . were reported killed or missing:* Estimates of deaths went as high as thirty. Various newspapers published different lists of the dead or missing. The only firm accounts of deaths are of the firefighters.

95 *"I got out my father's old revolver":* Murdoch, *Gentleman in Boston,* p. 119.

96 *"The number of tipsy men":* Conwell, p. 303.

96 *"What's the use of having"*: Murdoch, *Gnetleman in Boston*, p. 120.
96 *"The sun went down"*: Henry Ward Beecher, cited in *Chandler & Co.'s Full Account of the Great Fire.*
97 *"a very self-satisfied lot"*: *Daily Patriot*, Concord, New Hampshire, November 14, 1872.
97 *"hardly had time to take a breath"*: J.W. Black, *The Philadelphia Photographer*, Vol. 9, No. 108, December 1872, cited by Novack, *Photography and the Great Fire of November 1872*, title page.
97 *"Well, it's gone"*: Patrick Donahoe, *Chandler & Co.'s Full Account of the Great Fire.*
98 *"True, there were numerous instances"*: *Chandler & Co.'s Full Account of the Great Fire.*
98 *The mayor of Chicago:* Frothingham, p. 14.
98 *Harriet Beecher Stowe contributed:* Conwell, p. 290.
98 *"It would have been well"*: *Chandler & Co.'s Full Account of the Great Fire*, p. 46.
98 *"There don't seem to be any sick horses around"*: *Saturday Evening Express*, November 16, 1872.
99 *"I have been to oakum fires"*: Damrell's testimony, *Report of the Commissioners*, p. 114
100 *"This danger had been foreseen"*: *Report of the Commissioners*, p. vi.
101 *"brought into use many devices"*: Charles S. Damrell, "How the Fire Department Met the Crisis," *Current Affairs*, October 30, 1922, p. 32.
101 *"Could it happen again?"* Interview with John Vahey.
102 *"John Damrell, though castigated"*: See resolutions passed by a convention of fire engineers on November 22, 1872, which stated "that the ability, coolness, indomitable courage and perseverance displayed by Chief Engineer John S. Damrell . . . merits our . . . earnest and unqualified approbation." The engineers also called the use of gunpowder ineffectual. Brayley, p. 283.
102 *"He has been conceded"*: *Boston Transcript*, November 4, 1905.

CHAPTER 6 TWICE BURNED IN CHELSEA

103 *"Notify Newton Control"*: Interview with Herbert C. Fothergill, December 2002.
104 Background on the 1908 fire: Lyons, *Fire in America*; Chelsea Fire Department 1908 fire report, April 12, 1908.
104 *"as dead as Chelsea"*: Pratt, p. 15.
105 *"The consequence was"*: The Underwriters' Bureau of New England, "Report No. 118 on the Chelsea Conflagration of April 12, 1908," p. 5.
106 *One of the last companies to leave their post:* Charles F. Haywood tells this story of the Lynn fire department in *General Alarm*, pp. 63–65.
107 *"It seemed as if everyone tried first:"* Pratt, p. 44.
107 *The parade of people and possessions:* Ibid., pp. 55–58.
108 *In the People's AME Church: Boston Advertiser*, April 13, 1908, scrapbook in Drukman collection, Bostonian Society.
108 *"The flames seemed determined"*: unidentified newspaper clipping, "Flames Cut Swath Through Chelsea, scrapbook, collection of Herbert Fothergill.
108 *"The fire brooked no interferences"*: Ibid.
109 *"The wails of hundreds of parents"*: Pratt, p. 64.
109 *Eli C. Bliss:* Ibid., pp. 58–59.
112 *President Theodore Roosevelt telegraphed:* Ibid, p. 104.
114 *"The residents of Chelsea"*: Ibid., p. 148.
114 *Fire Chief Herbert Fothergill:* Interview with Fothergill.

116   *Eyewitness accounts of the 1973 fire:* December 2002 interviews with former Chelsea chief Fothergill and retired firefighters William Coyne and Leo Graves. Background information from *Fire Command!,* December 1973; Henry T. Hanson, "Conflagration Rages Through Chelsea Mass," *Fire Engineering,* December 1973; "The Great Chelsea Fire of 1973," special section, *Chelsea Record,* October 28, 1998; special section, *Boston Herald,* October 21, 1973; and Michael D. Drukman, *Conflagration! A Report on the Chelsea, Massachusetts Fire of October 14, 1973* (self-published); Herbert C. Fothergill, "Chelsea Revisited," paper for the Department of Fire Services, October 24, 2000.

120   *"The flames seemed to have an unusual intensity":* Drukman, p. 13.

124   *"If they're going to kill you for a truck":* interview with Fothergill; also reported in *Chelsea Record,* October 28, 1998.

CHAPTER 7 THE COCOANUT GROVE

Primary sources for this chapter include interviews with Charles Kenney (April and October 2002), Jack Deady (November 2002), Martin Sheridan (October 2002), John Quinn (October 2002), Jane Alpert Bouvier (January 2003), Paddy Noonan (October 2002), and Paul Christian (July and November 2002); also Benzaquin, *Fire in Boston's Cocoanut Grove;* Keyes, *Cocoanut Grove;* Lyons, *Fire in America;* Austen Lake, *Galley Slave* (Boston: Burdette and Company, 1965); John Vahey's analysis of the Cocoanut Grove fire, "Design for Disaster" (Boston: Boston Sparks Association, 1982); Robert Moulton, "Looking Back at the Cocoanut Grove," NFPA Report, *Fire Journal,* November 1982.

127   *Veteran firefighter Charles Kenney:* Interview with son, retired Boston firefighter Charles C. Kenney.

128   *In East Boston . . . John F. Crowley:* Details on Crowley's experiences are from his letter to Charles Kenney, August 18, 1991, provided courtesy of Kenney and the Crowley family.

129   *How did a club like the Cocoanut Grove:* The history of the Cocoanut Grove is based on Keyes, pp. 174–188; Benzaquin; articles by Austen in the *Boston Record* and *Boston American,* December 14–31, 1941; and a memorial pictorial section on the fire, *Boston Sunday Advertiser,* December 6, 1942.

130   *"Maybe he prints his own money":* Lake, *Galley Slave,* p. 223.

130   Details on Charles "King" Solomon: Lake, *Boston Record* and *Boston American* articles.

131   *"The dirty rats got me": Boston Post,* May 12, 1934.

132   *"the bland, monosyllabic young lawyer":* Lake, *Boston Evening American,* December 17, 1942.

132   *"typical American success story":* "Barnett Welansky Unmoved by Verdict," *Boston Globe,* April 11, 1943.

133   *"Mayor Tobin and I fit":* Testimony by Henry Weene before Reilly commission, December 1942. Tobin promptly issued a statement of denial.

134   *Everyone in Boston knew Marty Sheridan:* Details of Sheridan's experience from interview with Sheridan, and from his manuscripts "I Am Living on Borrowed Time" and "I Survived the Cocoanut Grove Disaster."

135   *Monogram Pictures:* This company eventually evolved into Allied Artists.

137   *Seventeen-year-old Dotty Myles:* Details of Myles's life are from interview with Charles Kenney and from Dorothy Myles, "I Almost Burned to Death," *True Experiences,* October 1948.

137 *Twenty-one-year-old John Quinn:* Details of his experience are based on interview with Quinn; Quinn's personal account in "Inferno at the Cocoanut Grove," *Yankee,* November 1998, pp. 125–126; letter to Bostonian Society, vertical file.

138 *it felt curiously warm:* Historians of the fire are puzzled by Sheridan's recollection; it does not seem to fit any known theory of how the fire started.

139 *"No one leaves until they pay":* Benzaquin, p. 34.

140 *Behind the bar:* Details of Daniel Weiss's experiences are from Benzaquin, p. 58, and from Judy Bass, "No Way Out," *Boston Magazine,* October 1992, p. 79.

143 *Saxophonist Al Willet:* Bass, pp. 79, 108.

145 Details of the response of the Boston Fire Department: Accounts in Benzaquin and Keyes; Boston Fire Department records; interview with Charles C. Kenney.

148 *Coast Guardsman Clifford Johnson:* Johnson's story is told in great depth in Benzaquin, pp. 153–184, and in Keyes, pp. 247–277.

150 *Club staffers even helped Lieutenant Murphy:* For Murphy's recollections, see "30th Anniversary of the Cocoanut Grove," *Boston Advertiser,* November 23, 1972.

150 *"Collins couldn't figure out":* "Last Dance at the Cocoanut Grove," *NFPA Journal,* May/June 1991, p. 8. Firefighter George "Red" Graney also encountered bodies in strange death poses, as described in Bass, p. 105, and other accounts.

152 Descriptions of victims' appearance: Francis Moore, *A Miracle and a Privilege: Recounting a Half Century of Surgical Advances* (Washington, D.C.: Joseph Henry Press, 1995) pp. 62–63.

152 *"But their flowers aren't even burned":* Thomas H. Coleman, "A Hush on the Brick Corridor," *Harvard Medical Alumni Bulletin,* Winter 1991/92, p. 15.

152 *Dr. Francis Moore . . . saw a naval officer:* Moore, p. 62.

152 *a patient was arriving every eleven seconds:* Moulton, NFPA Report, p. 7.

153 *At MGH medical history was made:* Oliver Cope, "End of the Tannic Acid Era," *Harvard Medical Alumni Bulletin,* Winter 1991/92.

153 *Dr. Thomas Risley:* "Inferno at the Cocoanut Grove," *Yankee,* p. 126.

154 *While waiting for treatment:* Joe Derrane recalls Myles talking about this incident.

155 *Busboy Tony Marra:* Benzaquin, pp. 59–60.

157 *It was Al:* Bass, p. 108.

157 *A quirk saved Dr. Joseph Dreyfus:* Charles C. Kenney interview with Dreyfus.

157 *"At his funeral":* "Final Honors to Buck Jones," *Boston American,* December 8, 1942.

157 *a bit of Hollywood hype:* "Final Honors to Buck Jones," *Boston American,* December 8, 1942.

160 *the six companies . . . paid out only $22,420:* Marty Sheridan, "Repercussions of Grove Fire Still Being Heard in Courts," *Boston Globe,* November 28, 1943.

160 Details on Philip Deady: Interview with Deady's son, Jack Deady.

160 *"The fire has revealed a weakness":* "Pastors Blast Officials in Fire" *Boston Evening American,* December 7, 1942.

160 *John "Knocko" McCormack:* interview with Jack Deady.

161 Quotes from Callahan and Doyle: "Jury Will Debate Grove Case," *Boston Record,* April 10, 1943, and other contemporary newspaper accounts of the trial.

164 *"suffered the tortures of the damned":* Charles C. Kenney, "Cocoanut Grove Busboy's Legacy of Pain and Suffering," *Middlesex News,* December 3, 1995.

164 *"Much of the cloth"*: William Arthur Reilly, fire commissioner, "Report concerning the Cocoanut Grove," November 28, 1942, p. 43.

164 *"This fire will be entered"*: Ibid, p. 48.

164 *"After exhaustive study"*: Official finding by State Fire Marshal Stephen C. Garrity, November 8, 1943.

165 *Nazi saboteurs:* Lake, *Galley Slave,* p. 217.

166 Methyl chloride theory: Interview with Charles C. Kenney.

166 *In 1996 Doug Beller:* Doug Beller and Jennifer Sapochetti, "Searching for Answers to the Cocoanut Grove Fire of 1942," *NFPA Journal,* May/June 2000.

168 *The classic example:* Benzaquin, pp. 149–152; Keyes, p. 259.

168 *"I want a mirror"*: Myles, "I Almost Burned to Death," p. 56.

170 *"There's nothing to worry about"*: Ibid., p. 58.

172 *"people must come out of bad experiences"*: "After 18 Years Singer Recalls Cry of 'Fire' at Cocoanut Grove," *New York World Telegram and Sun,* December 27, 1960.

173 *After the fire Mickey Alpert:* Interview with Alpert's daughter, Jane Alpert Bouvier.

173 *Timothy Leary:* The medical examiner who examined Cocoanut Grove victims was also named Timothy Leary.

174 *Fifty years after the fire:* Sandy Coleman, "Tragedy is commemorated: Ceremony at Cocoanut Grove site sees bittersweet reunion," *Boston Globe,* November 29, 1992; Joe Heaney, "Cocoanut Grove 50-year ceremony set," *Boston Herald,* November 27, 1992.

CHAPTER 8 DEATH IN THE VENDOME

176 Eyewitness accounts from interview with retired firefighter James McCabe, January 2003. Background information from an interview with Richard Magee, cited in "The Day Boston Cried," by Steve Buckley, *Yankee,* June 1997.

178 Background on the Vendome building and firefighting response from "Without Warning: A Report on the Hotel Vendome Fire June 17, 1972," by John Vahey, district fire chief, from Boston Fire Department archives; "The Vendome Fire," NFPA Fire Record Department, September 1972.

180 *One early arrival was Richard B. Magee:* Interview with Magee's son, Richard B. Magee, district fire chief, January 2003.

181 *Another early arriver:* Buckley, p. 68.

181 *"there were shouts of encouragement"*: Tom Berube, "Firefighter Never Doubted Rescue," *Record American,* June 19, 1972.

184 *A department chaplain:* Buckley, p. 68.

184 *Frances Dolan wept:* "City Mourns Firefighters," *Record American,* June 19, 1972.

184 *"He never got over the Trumbull Street fire"*: Ibid.

187 *"I embrace this"*: John Damrell letter, courtesy of Commissioner Paul Christian.

188 Background on Luongo Restaurant and Trumbull Street fires: Boston Fire Department Web site, www.ci.boston.ma.us/bfd/.

188 *"Depends on your point of view"*: Interview with Leo Stapleton, December 2002.

CHAPTER 9 MR. FLARE AND THE RING OF FIRE

192 *As early as 1652:* Brayley, p. 6.

192 *gang of "fire bugs"*: Ibid., p. 14.

192 *"Ah Boston"*: Cannon, p. 59.

192 *his ears were lopped off*: Ibid., p. 60.

192 *"A Negro woman"*: Brayley, p. 21; Seaburg, p. 123.

193 *"Red Roosters"*: various articles in *Boston Herald* and *Boston Post,* April 15, 1916.

193 Background on the Symphony Road fires: Harvey Schmidt, "'Boston's Neighborhood Arson Fighters," *International Association of Fire Chiefs,* April 1978, p. 6; Steve Slade, "Arson: Business by Other Means," *The Nation,* October 1977, p. 307; Joseph P. Blank, "They're Burning Our Neighborhood!" *The Reader's Digest,* October 1978, p. 131; James Brady, "Arson Urban Economy, and Organized Crime: The Case of Boston," *Social Problems,* October 1983, p. 1–27.

194 *Mortgage-holding banks:* Mark Zanger, "South Boston Saving Bank Fiddles While Boston Burns," *Real Paper,* April 22, 1978.

194 *"Maybe an owner does hire a torch"*: Slade, p. 308; Brady, p. 18.

194 *"initially most people refused to believe"*: Cited by Brady, p. 18.

195 Background on First Security Services and subsequent legal action: Interview with Larry Curran (January 2003); James M. Connolly, "Busted in Boston," *Firehouse,* August 1982, p. 33.

196 *Boston, a city with a long tradition:* Michael Blanding, "Some Like It Hot," *Boston Magazine,* March 2001; Harry Belknap, "The Box 52 Association: A Unique Boston Organization," *Fire and Water Engineering,* December 3, 1924; Paul Ditzel, "USA's No. 1 Fire Fan, " *Firehouse,* May 1983; interviews with Ted Gerber (July and November 2002), Frank W. Fitzgerald Jr. (March 2003), and Elliot Belin (January 2003).

196 *"that great fraternity"*: "I Discovered the Chelsea Fire," undated newpaper clipping from Herbert Fothergill collection.

197 *"Suddenly, a huge sheet of flame"*: "'Huge Flames' Belched 15 Feet to Street," *Boston Globe,* November 29, 1942.

197 *"Fires are fascinating"*: Paul Benzaquin, "Fire Buff Ben Ellis Was at Scene," *Boston Globe,* January 13, 1960; also Lawrence Dame, "'Spark' Working At Hobby Is Happy Citizen," *Boston Herald,* October17, 1948.

198 Details on Gregg Bemis's life: Christopher Whipple and Enrico Ferarelli, "Fire Starters: The Strange Tale of the Men Who Tried to Burn Down Boston," *Life,* April 1985; William F. Doherty, "Man Accused of Arson Says Fires Helped Tregor Bill Pass," *Boston Globe,* November 15, 1984.

199 Details on the 1980s arson fires and background on the suspects: Interviews with Wayne Miller (February 2003), Stephen McLaughlin (February 2003), Elliot Belin (January 2003), and Leo Stapleton (December 2002); "A Ring of Fire," *Boston Magazine,* November, 1984; "Who's Burning Boston?" *Boston Magazine,* December 1982; Paul Teague "Big Bust in Beantown," *Firehouse,* October, 1984; "Boston Fires an Omen for All Cities," *US News and World Report,* February 7, 1983; contemporary news articles in the *Boston Globe* and *Boston Herald.*

201 *"At first, it did not appear"*: Stapleton, *Commish,* p. 10.

202 *the work of vandals or thrill-seeking juveniles:* Jahnke, "Who's Burning"; Frank Mahoney, "Comr. Paul Fears Major Arson Problem Developing," *Boston Globe,* August 22, 1982.

203 *"We were going crazy"*: Interview with Wayne Miller.

205 *fire destroyed the Fourth Street headquarters:* Paul Ditzel, "Boston Sparks Burned Out," *Firehouse,* May 1983, p. 56.

205 *"the light came on slow"*: Interview with Stephen McLaughlin.

208 *the judge said he was puzzled:* Andrea Estes, "Officer Gets 12 Yrs. in Hub Arson Spree," *Boston Herald,* July 7, 1984.

208 *"I was naturally bitter":* Paul Teague, "A Victim of the Torches," *Firehouse,* October, 1984, p. 2.

209 *"Every little one":* Interview with Stephen McLaughlin.

209 *"it was a fantastic fire":* Bemis's testimony, cited in Beverly Ford, "Accused Torch Bragged of $13.5 M Fire: Witness," *Boston Herald,* November 10, 1984.

210 *Stackpole was found guilty:* Two other men were charged in connection with the conspiracy: Mark Svendbye received three years in prison and Chris Damon received five years probation, according to law enforcement officials.

210 *"These guys acted like fire was a football game":* Interview with Elliot Belin.

ILLUSTRATION CREDITS

Images in jacket montage: top, Chelsea in flames, from the collection of Herbert C. Fothergill, courtesy of Mr. Fothergill; lower right, rescue from the Cocoanut Grove, from the collection of Bill Noonan, courtesy of Mr. Noonan, the Boston Public Library, and the *Boston Herald*; lower left, firefighters at the Vendome Hotel fire, Kevin Cole photo, courtesy of the *Boston Herald.* Image on back of jacket: "Boston in Flames," Currier & Ives lithograph of the Great Fire of 1872, courtesy of The Bostonian Society/Old State House.

Endpapers illustration by Jeffrey M. Walsh; pp. vi, 11, courtesy of Boston Public Library; pp. 7, 31, 43, 73, 79, 88, 89, 90, 91, courtesy of The Bostonian Society/Old State House; pp. 8, 37, courtesy, CIGNA Museum and Art Collection; p. 14, courtesy of American Antiquarian Society; pp. 17, 22, 27, courtesy of Nancy Lusignan Schultz; p. 38, from the collection of Edward R. Tufts, courtesy of Mr. Tufts; p. 41, from the collection of J. Richard Hunneman, Jr., courtesy of Mr. Hunneman; pp. 49, 59, 63, *The Fireman's Herald,* 3 April 1902, courtesy of the Massachusetts Historical Society; pp. 65, 99, 172, 189, 190, photos by Stephanie Schorow; pp. 68, 76, 92, Russell Conwell, *History of the Great Fire in Boston,* 1873; p. 81, 95, "Carleton," *The Story of the Great Fire,* 1872; pp. 111, 112, postcards from the collection of the Stephanie Schorow; p. 110, Walter Merriam Pratt, *The Burning of Chelsea,* 1908, image provided courtesy of the Somerville (Mass.) Public Library; p. 113, courtesy of the *Boston Herald;* pp. 118, 122, 124, from the collection of Herbert C. Fothergill, courtesy of Mr. Fothergill; p. 123, Stanley Samson photo, courtesy of the *Boston Herald;* p. 133, from the collection of Kathy Alpert, courtesy of Ms. Alpert; p. 133 inset, from the collection of Jack Deady, courtesy of Mr. Deady; p. 135, from the collection of Martin Sheridan, courtesy of Mr. Sheridan; p. 144, courtesy of the *Boston Herald,* Bill Noonan, and the Boston Public Library; pp. 145, 146, 149, 180, 183, 201, 206, from the collection of Bill Noonan, courtesy of Mr. Noonan; p. 156, copyright, 1942, the Globe Newspaper Co.; republished with permission of Globe Newspaper Company, Inc.; p. 158, illustration by Jeffrey M. Walsh; p. 162, courtesy of Bill Noonan and the Boston Public Library; p. 165, courtesy of AP/Wide World Photo; p. 171, from the collection of Charles Kenney, courtesy of Mr. Kenney; p. 185, Kevin Cole photo, courtesy of the *Boston Herald;* p. 204, Joanne Rath photo, courtesy of the *Boston Herald;* p. 209, Jim Mahoney photo, courtesy of the *Boston Herald.*

INDEX

Italic page numbers refer to illustrations.

Adams, Samuel, 11, 12, 92
Alarm(s). See Fire alarm(s); Fire alarm
    boxes
"All out" (code 22–22), 66, 124
Alpert, George, 129–30, 173
Alpert, Mickey (Milton Irving), 129,
    134, 151, 157, 159, 173
American Fire Alarm Telegraph, 56–57,
    61. See also Fire alarm system
Amoskeag Manufacturing Company,
    43, 46
    Kearsarge No. 3, 80, 91
Amsterdam, Netherlands, 8
Animal Rescue League, and 1908
    Chelsea fire, 114
Annexation, and fire alarm system, 59,
    66
Anti–Catholic bias. See Catholicism
Anti-Irish sentiment, and Ursuline
    Convent fire of 1834, 13, 18, 23
    Broad Street Riot and, 25–26
    criminal convictions and, 26
    fire department membership and, 40
    See also Irish immigrants
Antismoking rules, 6
Arson
    in 1916, 193
    in 1970s and 1980s, 191–211
    cause of 1679 fire, 6, 192
    complaints of STOP, 194–95, 196
    in early Boston, 6, 192–93
    in Fenway neighborhood, 193–96
    and firefighters injured, 200, 204–5,
      208, 210
    for-profit, 191, 193–96
    increase in 1982, 200–2
    investigation into, 195–96, 202–8
    methods for, 196, 202
    motives for, 193, 196

    Proposition 2½ cutbacks and, 200
    See also Arsonist(s); "Mr. Flare"
Arsonist(s)
    Fraine, Francis ("Frankie Flame"),
      195–96
    Lincoln, George W., 195
    photographers' rule for, 193, 196,
      205–6
    "Red Roosters," 193
    See also "Mr. Flare"
Atkins, Thomas, 8
Auxiliary firefighters, 123, 198, 199

Bean and Scott, 43
Beck, Chelsea Mayor John E., 112
Beecher, Rev. Lyman, 15, 18, 89, 97
Belin, Elliot ("Captain Flame"), 198,
    210–11
Bell, Alexander Graham, 62, 63, 64
Bellflower Street Fire (Dorchester), 215
Bellotti, Mass. Atty. Gen. Francis X.,
    194, 195, 196
Bemis, Gregg, 198–200, 203, 205, 206,
    208, *209*, 210
Benham, Gen. Henry W., 86, 88–89
Benzaquin, Paul, 144
Berman, Jack, 129, 130
Bodenhorn, Reuben, 129, 157, 161
Boston's Common Council, 69
Boston Blacking Company, 105
Boston City Hospital (BCH), 144, 148,
    152, 169, 170
Boston Common
    fire of 1872 and, 77–78, 96
    musters at, 39, *40*
Boston Fire Department
    Cocoanut Grove Fire and, 126, 160
    established, 12
    most devastating loss at Vendome, 84

Boston Fire Department (*cont.*)
  rapid response system of, 49
  See also specific fires and fire personnel
Boston Fire Museum, 211
Boston Fire Society, 8, 9
Boston Globe, 155, *156*, 168
Boston Herald, 155, 213
Bostonian Society, 62n
Boston Pilot Building, *79*
Boston Sparks Association, 197, 200,
  205, *206*, 210–11
Boston Town House, 3
Brayley, Arthur W., 6, 25, 26, 192
Brazen Head, site of 1760 fire, 10
Broad Street Riot of 1837, 24–29, 39.
    See also Ursuline Convent
Brooks, Rev. Phillips, 82n, 89, 90–91
Bucket brigades, 2, 12, 34, 36, 41
Building codes, 101, 166, 167, 216. *See
    also* Rules and regulations
Bulfinch, Charles, 12
Burning of Chelsea, The (Pratt), 107
Burning of the Convent, The (Whitney), 28
Burn treatment, 153, 163
Burt, William L., 85–86, 92, 94, 101
Bushnell, Mass. Atty. Gen. Robert, 159,
  161, 162
Buzzell, John R., 16, 19, 20, 21, 23–24, 27

Callahan, Herbert F., 132, 161
Cannon, Dr. Bradford, 153
Capistran, Dep. Chief Bill "Cappy,"
  119, 125
Cathedral of the Holy Cross, 28
Catholic archdiocese, 24
Catholic Cathedral (Boston), 23
Catholicism
  biases against, 13–15, 18, 24
  sermons against, 89
Channing, William Ellery, 51
Channing, William Francis, 48, *49*
  death of, 64
  life ambitions of, 50–51, 60
  See also Farmer, Moses Gerrish; Fire
    alarm system; Gamewell
Charitable Fire Society, 32
Charlestown, Massachusetts, 15–23, 29
Chelsea Fire of 1908, 104–115, 120,
  121, 125
  Box 215, 105

buildings burned during, *111, 112, 113*
cause unknown, 114
deaths resulting from, 104, 112
estimated damages from, 104, 112
oil hazard, 105, 109–10
and out-of-town firefighters, 106–7
"rag district," fire risk, 105–6
railroad tracks, as fire break, 109, 111
Spencer, Fire Chief Henry A., 104,
  105, *110*, 111
See also Pratt, Walter Merriam
Chelsea Fire of 1973, 103–4, 116–25
  Box 215, 116
  Capistran, Dep. Chief Bill "Cappy,"
    119, 125
  cause unknown, 124
  City Hall, 120, 121, 125
  Coyne, Dep. Chief William, 116–17,
    119, 120–23, 125
  estimated damages from, 124–25
  fire station, 121, 125
  Fothergill, Fire Chief Herbert C.,
    103–4, 114–15, 117, 119–25, *122*
  Fothergill's car, 117, 120–21, *124*
  Graves, Fireman Leo, 119, 121, 122,
    125
  low water pressure in, 120–22
  multiple alarms for, 116–17
  oil hazard, 120
  and out-of-town firefighters, 117,
    121, 122–23
  same path as 1908 fire, 120
  Williams School in, 120, 124, 125
Chicago fire of 1871, 67, 69–70, 75, 80,
  94, 97, 98, 102
Chimney fires, in early Boston, 4
Church bells, as fire alarms, 49, 54, 60
Cisterns, 12
Claflin, Gov. William, 63
Claflin Guards, 94
"Cocoanut Grove Law," 167
Cocoanut Grove Nightclub Fire,
    126–75
  alarms for, 128, 147
  ball of flame, 141, 159
  Box 1521, 146, 147
  building layout, 133–34, *158*
  cause debated, 126, 139, 164–67
  and combustible furnishings, 129,
    133, 139, 140, 163–64

Crowley, Dist. Chief Daniel, 128,
    145, 147
deaths resulting from, 143–44,
    146–47, 150–52, 155, 157, 159, 168
faulty wiring and, 133, 164, 165, 166
and firefighters, 144–48, 150–51
and football game patrons, 127
hospitals involved in, 144, 148,
    152–53, 163, 167, 169, 170
indictments, 161
investigation into, 159–60
items not burned, *149*, 164–65
lights out during, 140, 141
and locked exit doors, 131, 139, 141,
    143, 162, 166
Melody Lounge as origin of, 139,
    140
memorial plaque for, 174
occupancy limit and, 134, 138
panic during, 142–43, 155, 161–62
Prohibition and, 127, 129, 130, 131,
    134
pungent smoke odor in, 141, 143,
    150, 163
remodeling prior to, 129–30
rescues, *144*, 146–48, 151, 197
revolving doors and, 139–40, 141,
    143, 166
rollback roof and, 130, 150
route of, 139, 144, 159
Solomon, Charles "King," 130–31
speed of, 139, 144, 165
streets abutting, 134, 141, 142, 143,
    146–48, 150
survivor stories, 140–42, 153–54,
    155, 156–57, 159
trial for, 160–63
underworld and, 131, 132, 160–61, 167
See also Alpert, Mickey; Cocoanut
    Grove survivors; Cocoanut Grove
    victims; Jones, Buck; Kenney,
    Charles; Myles, Dotty; Sheridan,
    Martin; Tomaszewski, Stanley;
    Welansky, Barnett
Cocoanut Grove survivors, 140–42,
    153–54, 155, 156–57
Dreyfus, Joseph, 137, 157
Dunlap, Scott, 153–54, 172
Johnson, Clifford, 148, 170–71
Kelley, Joseph F., 159

Kelley, Joseph F., 159, 166
Luby, Marion, 138, 140, 170
Marra, Tony, 155–56, 174
Quinn, John, 137–39, 140, 141–42, 170
Siegel, Henrietta, 157, 172
Vient, Dick, 137–38, 140, 170
Weiss, Daniel, 140, 150
Whitehead, Gerry, 137–38, 140,
    141–42
Willet, Al, 156–57, 172
See also Kenney, Charles; Myles,
    Dotty; Sheridan, Martin
Cocoanut Grove victims
Andrews, Lynn, 155
Balzarini, Frank, 155
Dreyfus, Adele, 161
Fazioli, Bernie, 155
Fitzgerald, four sons of Mary, 155
Gatturna, Francis, 168
Gebhard, Charles (Buck Jones), *135,*
    135–37, 153, 154, 157–58
Goodelle, Goody, 138–39, 155
Johnson, Clifford, 148, 170–171
Jones, Buck, 135–37, 151, 153, 154,
    157, 159
O'Neill, John and Claudia, 137, 155
servicemen, 155
Shea, Tiny, 138, 155
Sheridan, Connie (Mrs. Martin), 137,
    141, 168
Swett, Katherine, 133, 150, 155
Viator, Stanley, 148
WAVEs, 155
Coffin, Charles Carleton, 89, 93
Commish (Stapleton), 201
Competitions. *See* Musters
Conflagration, 103, 105, 215, 216
    definition of, 103
    See also Chelsea Fire of 1908;
        Chelsea Fire of 1973; Fire of 1760;
        Great Fire of 1872
Contests. *See* Musters
Convent fire of 1834. *See* Ursuline
    Convent
Conwell, Russell H., 73
Cope, Dr. Oliver, 153, 163
Copley, John Singleton, 91
Cornhill Row, 1–2, 10
Coyne, Dep. Chief William, 116–17,
    119, 120–22, 125

Crane, Moses, 58
Crowley, Dist. Chief Daniel, 128, 145, 147
Crowley, Gerard, 214
Crowley, John F., 128, 129, 150–51, 174, 129, 214
Curran, Larry, 195, 196
Currier & Ives, 32, 79
Currier, Nathaniel, 31, 32
Cutter, Edward, 17, 22

Damrell, Charles, 101
Damrell, Chief Eng. John Stanhope
compared to Herbert Fothergill, 115
    and decision to use explosives, 75–77, 85–90, 98, 102n
    directing firefighting, 67–72, 74–75, 80, 82–88, 92
    Engine Company No. 11, 43, 69
    after the fire, 99–102
    fire helmets of, 99, 102, 213
    and note to fireman's widow, 187
    testimony of, 212–13
Dana, David D., 12, 39–40
De Furia, State Police Lt. James, 194, 196
Delano, William E., 84
Ditzel, Paul, 4, 5, 6, 41
Dolan, Fire Marshal Joseph, 194
Dover, Massachusetts, 102, 213
Downtown Crossing, 102
Doyle, Asst. D.A. Frederick T., 162
"Dressed up like a fire engine," 30
Drukman, Michael, 120
Dunbar, Capt. Joseph, 88

East Boston, fire in Luongo Restaurant, 128, 188
Electromagnetism, 53
Eliot, Charles, 82, 83
Eliot, Mayor Samuel, 26–27, 83
Ellery, William, 50
Ellis, Ben, 197
Emerson, Ralph Waldo, 51
"Epizootic," and the Great Fire of 1872, 67, 71, 74, 98–99, 102
Explosives, and the Great Fire of 1872, 75, 77, 85–90, 98, 100, 102n

FAIR plan, 195
Faneuil Hall, 6, 23, 60, 85

Farmer, Moses Gerrish, 48, 49, 53–64
    death of, 64
    and electrical engineering, 63
    inventions of, 53, 54–55, 63
    as mechanical genius, 48, 53
    See also Channing, William F.; Fire alarm system; Gamewell
Fenwick, Bp. Benedict Joseph, 15
Fiedler, Arthur, 197, 198
Fire, "working," 176, 179
Fire alarm(s)
    colonial methods, 49, 60
    multiple, 60
    See also Fire alarm boxes
Fire alarm boxes, 59, 65,
    Box 52, 67, 100, 196–97
    Box 215, 105, 116
    Box 1521, 146, 147
    in Chelsea conflagrations, 116
    color of, 55, 60, 101
    as heart of response system, 48–49
    instructions inside, 55–56
    keyless, 60, 101
    locked, 56, 72, 74
    mechanics of, 54–56, 60, 65
    numbered, 49, 54, 59, 66
    "phantom," 66
    stolen, 203
    telephoned reports and, 65
Fire alarm system, 48–66, 57
    in Boston Fire Museum, 211
    compensation for invention, 56, 61
    computers and, 64–65
    cost of original system, 55
    first municipal, 48
    installation of original system, 55
    office, 60, 63, 65
    patent for, 56, 57, 58, 61
    recent enhancements, 64–65
    sales of, 58
    See also Farmer, Moses Gerrish; Channing, William Francis; Gamewell
Fire and Roses (Schultz), 18–19
Fire boxes. See Fire alarm boxes
"Fire bugs," 6, 192. See also Arsonist(s)
Fire clubs
    Boston Sparks Assn., 197, 200, 205, 206, 211
    Box 52 Assn., 196–97

Tapper Club, 197, 211
See also "Sparks"
Fire codes, 167, 216. *See also* Rules and regulations
Fire districts, 11, 49
creation of, in early Boston, 11
as part of fire alarm system, 54, 59
"Fire engine red," 32
Fire engines
decoration of, 30–32, 36, *37*, 43, 46
preservation of old, 47
response time and, 37–38
See also Hand pumpers; Hunneman fire engines
Firefighters
auxiliary, 123, 198, 199
first time paid, 8
and layoffs in 1981, 200
professional, 44, 60
See also Volunteer firefighting; *specific fires and firefighters*
Firefighting innovations
aversion to, 34, 41
harness preparation, 44
hoses, 41, 46–47
keyless alarm system, 101
red alarm boxes, 101
suction device, 36
telegraphy, 48
two-way radios, 60, 103–4
wagon for hoses, 36
See also Fire alarm boxes; Fire alarm system
Fire horses. *See* Horses
Fire of 1653, 5, 34
Fire of 1676, 6, 34
Fire of 1679, 6, 192
Fire of 1711, 2–3, 6, 8, 9
Fire of 1760, 10–11
Fire of 1872. *See* Great Fire of 1872
Fire prevention, 9, 12. *See also* Building codes; Rules and regulations
Firestorms, in Chelsea, 106, 109, 119
Fire wards, in early Boston, 11,12
First Security Services, 195
Fothergill, Fire Capt. Herbert, Sr., 114–15
Fothergill, Fire Chief Herbert C., 103–4, 114–15, 117, 119–25, *122*
Fraine, Francis ("Frankie Flame"), 195–96

Franklin, Benjamin, 9
"Friday Firebug," 202, 203

Gamewell Company, 61, 65
Gamewell, Frank Asbury, 58
Gamewell, John Nelson, 57–59, 61
Gamewell, Kennard & Co., 58
Gardiner, James M., 59
Garrity, Fire Marshal Stephen, 165
Gaston, Mayor William, *68,* 84–85, 86
Gebhard, Charles. *See* Jones, Buck
Goddard, Louisa. *See* Whitney, Louisa Goddard
Graves, Fireman Leo, 119, 121, 122, 125
Great Fire of 1872, 67–102
advertisements after, *95,* 97–98
alarms triggered for, 74
Boston Common in, 77–78, 96
Box 52, 67, 101
Brooks, Rev. Phillips, 82n, 89, 90–91
Burt, William L., 85–86, 92, 94, 101
business affected by, 72–73
cause unknown, 73, 99
and Chelsea blaze of 1908, 105
Claflin Guards in, 94
damage estimates for, 94
deaths caused by, 82, 94, 126
districts affected by, 72–73, 88, 94, 97, 102, 212
Eliot, Charles, 83
explosives used in, 75–77, 85–90, 98, 102n
final report on, 99–100
firefighter deaths in, 82
Franklin St., *76*
gas supply and, 72, 78, 87, 91–92, 102
Gaston, Mayor William, *68,* 84–85, 86
historical plaques for, 100
Holmes, Dr. Oliver Wendell, 81, 96
intoxication following, 96
Kearsarge No. 3 in, 80, 91
looting during, 82, 94
mansard roofs and, 73, 74, 77, 78–80, *79,* 97, 100
Old South Meeting House, *90,* 92–93, 96, 101
origin of, *73*
out-of-town firefighters in, 80, *91*
post office, 93, 101

Great Fire of 1872 (*cont.*)
   Putnam, Sarah, 72, 77, 78, 87, 94–95, 98
   Savage, Police Chief E. H., *68*, 95
   steamers used in, 92–93
   water system and, 70, 80, 99, 100, 101
   See also Damrell, John Stanhope
Greenberg, Amy S., 32
Green, Rev. John Singleton Copley, 89
Groblewski, Robert, 199, 200, 205, 207–8
Gunpowder. *See* Explosives

"Hallooing fire," 49
Hancock, John, 11
Hand pumpers
   Boston's last, 43
   manufacturers of, 34
   preservation of old, 47, 211
   "running with the machine" and, 43–44
   transition to steam and, 44–46
   See also Hunneman Fire Engine
"Hand tub" fire engine, 7, *8,* 34–35, 36–37, 41
Harrison, Elizabeth (Sister Mary John), 17–18, 19
Holmes, Dr. Oliver Wendell, 82–83, 96
Holmes, Oliver Wendell Sr., 51
Holzman, Robert S., 38
Hook and ladder companies, 46
Horse distemper. *See* Epizootic
Horses, 44–46
   steam engines and, 42, 44
   trained for smoke and fire, 71
   viewed as equals to firefighters, 44–45
Hose companies, 46–47
Hotel Vendome. *See* Vendome Hotel
Howard-Davis, 34
Hunneman and Company, 45
Hunneman Company. *See* William C. Hunneman Company
Hunneman fire engines, 30–47
   hand pumpers, 43, 45–46
   names and numbers of, 36, 45–46
   patent notice for, *33*
   prices for, 36–37
   quality and design of, 33–34, 45, 46
   "Somerville No. 1" (No. 678), *40, 45
   steamers, *40,* 45, 80

"St. Louis," No. 574, *37*
Torrent," 36, *38*
Hunneman, John C., *38*, 45
Hunneman, Joseph, 45, 46
Hunneman, J. Richard Jr., 47
Hunneman, Samuel Hewes, *38*, 45
Hunneman, William C., 32, 35, 45
Hunneman, William Jr., 36, 45

"Ingines," 2, 34
Intercity Mutual Aid radio, 117
Intercity Mutual Aid System, 103
International Association of Fire Chiefs, 101, 125
Irish immigrants, 14–15
Iroquois Theater (Chicago), 126
Ives, James Merritt, 32

Jacobs, Capt. John, 84, 87
Jeffers Company, 34
John S. Damrell Engine company No. 11, *43,* 69. *See also* Damrell, John Stanhope
Johnson, Clifford, 148, 170
John W. McCormack Post Office and Court House, 100
Jones, Buck (Charles Gebhard), *135,* 151, 153
   Boston tour of, 136–37
   death at MGH, 153, 154
   funeral for, 157, 159
   movie career of, 135–36

Kennard, John, 58, 59
Kenney, Charles (junior), 165–66, 172, 174
Kenney, Charles (senior), 127–28, 129, 145, 146–48, 165, 171
   hospitalized, 148, 169
   last fire call of, 129, 172
   legacy of, 174, 214
   mourning losses, 127–28

Ladder companies, 46
Lake, Austen, 130, 132, 161, 166
Lane Theological Seminary, 18
Leather buckets, 2
Leslie Company, 34
Lincoln, George W., 195
Linney, Fire Lt. Frank J., 134, 161

Lippi, Angelo, 130, 131, 132, 138
Long Wharf, 4, 10
Looting
    in early Boston fires, 3, 8, 9
    in Great Fire of 1872, 82, 94
Luongo Restaurant fire (East Boston),
    128, 188
Lynn, Massachusetts, 216

McLaughlin, Stephen E., 200, 202, *204,*
    205, 210
McManus, Dorothy. *See* Myles, Dotty
Magee, Dist. Fire Chief Richard B., 214
Mahoney, Dist. Chief William J., 155
Mansard roofs, 78–80
    of Boston Pilot Building, *79*
    as contributor to 1872 fire, 100
    flammability of, 74, 77
    material and character of, 97
    origin of, 73
    warnings about, 78–80, 100
Mary John, Sister (Elizabeth Harrison),
    17–18, 19
Massachusetts Charitable Fire Society,
    12
Massachusetts General Hospital
    (MGH), and Cocoanut Grove fire,
    152, 153, 163
    Cocoanut Grove victim statistics,
    152
    medical history made, 153, 163
Mather, Cotton, 1, 3, 5, 6, 192
Mather, Rev. Increase 1, 3, 6
Memorial Hall, 187–88, 211
Methyl chloride, and Cocoanut Grove
    fire, 166
Metzger, Dorothy. *See* Myles, Dotty
Miles Greenwood (Boston's first steam
    engine), 42
Miller, Wayne, 202, 203, 207
Moffatt, Mary Anne. See St. George,
    Sister Mary Edmond
Moore, Dr. Francis, 152, 153, 172
Morse, Mary, 1, 2, 3
Morse, Samuel, 48, 52, 53, 55
Mother Superior. *See* St. George, Sister
    Mary Edmond
Mount Benedict, 15, 16, 19, 27, 28, 217
"Mr. Flare," and arson in 1970s and
    1980s, 191–211

Bemis, Gregg, 198–200, 203, 205,
    206, 208, *209,* 210
    and Bureau of Alcohol, Tobacco and
    Firearms, 202, 203, 207
    "Friday Firebug," 202, 203
    and gang, 206–7, 209, 210
    Gorman, Joseph M., 199, 208, 209, 210
    Groblewski, Robert, 199, 200, 205,
    207–8
    investigator for, *204,* 205, 210
    Kendall, Leonard A., 208, 210
    locations hit by, 200, 203–5, 210
    Norton, Ray J., Jr. ("Crazy Ray"),
    199–200, 205, 210
    note from arsonist, 203, 210
    Sanden, Wayne S., 199–200, 205–10,
    209
    Stackpole, Donald F., 199–200,
    206–11
Mullen, John A., 40
Musters, 39, *40,* 42, 47
Myles, Dotty (Dorothy Metzger), 137,
    171
    Kenney family and, 169, 172
    at onset of Cocoanut Grove fire, 142
    recovery of, 154, 169–70
    rescue of, 147–8
    and singing career, 137, 169–70, 172,
    213
Mugar, David, 197

National Association of Fire Engineers,
    101
National Fire Protection Association
    (NFPA), 115, 164, 166, 186
Naumkeag Stream Cotton Company
    (Salem, Mass.), 216
New England States Veteran Fireman's
    League, 47
New York Fire Department, 214
North Meeting House, 3, 6
Norton, Ray J., Jr. ("Crazy Ray"),
    199–200, 205, 210
"Nunnery." *See* Ursuline Convent

Oakum, 2, 98
Old Meeting House, 3
Old South Meeting House, 89, *90,* 91,
    94, 96, 101, 218
O'Mara, Dep. Chief John, 182

Peabody, Elizabeth Palmer, 51
Perkins, Jacob, 35
"phantom boxes," 66
"plug ugly," origin of expression, 38
Post Office Square, 100
Potato Famine, 15
Pratt, Walter Merriam, 107, 109–12, 114
Prohibition, 127, 129, 130, 131, 134
Prudential Tower fire, 215–16
Putnam, J. Pickering, 97
Putnam, Sarah, 72, 77, 78, 87, 94–95,
    98

Quincy, Mayor Josiah, 12, 53
Quirk, Bobbie, 128

"Rattle Watch," 9
Red Lion Tavern, 6
"Red Roosters," 193
Reed, Rebecca, 16, 24, 27
Reilly, Commr. William Arthur, 157,
    160, 166
Relays, for delivering water, 39, 122
Renard, Jacques, 129
Revere, Paul, 12, 32, 35
Risley, Dr. Thomas, 153
Roach, Dep. Chief William, 121
Rogers, Edwin, 58
Romance of Firefighting, The (Holzman),
    38
Roosevelt, Pres. Theodore, 112
Rosin Storage, 106
Roxbury Fire Department, 38
Rules and regulations
    in early Bsoton, 5–6
    for pyrotechnics, 175
    See also Building codes
Running cards, 60, 128, 211
"Running with the machine," 30–32,
    36, 43–44, 71

Safety codes, 167. See also Rules and
    regulations
Salem, Mass., 216
Saltonstall, Gov. Leverett, 160
Sanden, Wayne S., 199–200, 205–10,
    209
Savage, E. H., 68, 95
Scanlon, Fire Chief Joseph, 216
Schultz, Nancy Lusignan, 18

Scondras, David, 194, 196
Seventeenth-century fire, 7
Sheridan, Gen. Philip Henry, 70, 75
Sheridan, Martin ("Marty"), 134–35,
    135, 136–37, 138, 140, 148,
    151,160, 172
    career return of, 168–69, 173
    hospitalization of, 154, 167–68
Shreve, Crump and Low, 83n, 91
Signal box, 59, 65. See also Fire alarm
    boxes
"Smoke-eaters," 40, 42
Solomon, Charles "King," 130–31
Somerville, Mass., 27, 28, 217
Sotherden, Howard E., 148, 168–69
South End, 189, 218
"Sparks," 197–99, 205, 210–11, 212, 218
    Belin, Elliot ("Captain Flame"), 198,
        210–11
    comparison to"tuna fleet", 198
    Ellis, Ben, 197
    famous, 197–98, 211
    firefighters' attitudes toward, 197
    origin of term, 196
    See also Fire clubs; "Mr. Flare"
Speaking trumpets, 49
Spelman, Chelsea Mayor Phil, 121
Spencer, Chelsea Fire Chief Henry A.,
    104, 105–6, 110, 111
Spero Toy Company, 200, 210
Sprinkler systems, 119–20, 216
Stackpole, Donald F., 199–200, 206–11
Stapleton, Capt. John V., 188
Stapleton, Fire Cmsr. Leo, 188, 201, 214
State House, 6
State Street, 1797 fire, 11
Station, The (rock club), 2003 fire at,
    174–75
Steam engines, 41–47
    fire of 1872 and, 92–93
    development of, 41
    first in Boston, 42
    horses and, 42, 43, 44
    Hunneman, 41, 45, 80
    Kearsarge No. 3, 80, 91
    opposition to, 41
    riot at introduction of, 34
    transition from hand-pumpers,
        44–46
    transition to gasoline-powered, 46

St. George, Sister Mary Edmond (Mary Anne Moffatt), 16, 19, 20, 21, 23, 24, 27
Stickel, Dep. Chief Louis, 147
Stowe, Harriet Beecher, 15, 98
Stuyvesant, Peter, 9

Telegraph, 50, 52–53, 55n, 62. *See also* Fire alarm system; Morse, Samuel
Thaxter, Lucy, 16, 19, 20, 22
Thayer Company, 34, 39, 45
Tobin, Boston Mayor Maurice, 163
Tomaszewski, Stanley
   lit match at Cocoanut Grove, 139
   stigmatized, 163
   testimony of, 164, *165*
"Torrent Six," *38*
Town House, 6
Tregor Bill, 205
Trinity Church, 96, 101
Trumbull Street fire (South End), 184–85, 188–90
Trumpets, 49
Tufts, Edward R., 33, 45

Underwriter's Bureau of New England, 105, 114
Union Fire Company, 9
Ursuline Convent, 13, 15–24, *17, 27*
   "escaping" nun story, 16–18, 20
   fire at, 20–24, *22*
   rioters at, 19–24
   rumors about, 16–20
   See also Broad Street Riot; St. George, Sister Mary Edmond; Whitney, Louisa Goddard

Vahey, Dist. Fire Chief John, 101, 179, 186, *190*
Vendome Hotel, 1973 fire at
   Box 1571, 179
   building collapse, 181–82, *183*
   cause unknown, 185
   faulty remodeling of hotel, 185–86
   firefighters at, 179–84, *185*), 214
   firefighters killed at, 184, 214

funeral service after, 184, 185
Keating, Msgr. James, 182–83
McCabe, Lt. James ("Midget"), 176–78, 182–83
Magee, Richard B., 177–78, 180–181, 184
memorial, 178, *189*, 190
O'Mara, Dep. Chief John, 182
origin of, 177, 179, *180*
Volunteer firefighting
   American origins of, 7–9, 49
   salaried, transition to, 27, 44, 60
   as spirit of America, 31–32

Walsh, John, 151
"Washed engine," 39
Water
   in Great Fire of 1872, 70, 80, 99, 100, 101
   low pressure in Chelsea fire, 119–20
   relays, 39, 122
   sprinkler systems, 119–20, 216
   See also Cisterns
Watson, Thomas, 62
Welansky, Barnett, 130–131, 132, 138, *162*
   owner of Cocoanut Grove, 132–33, 160
   sentencing and pardon of, 163
   trial of, 161, 162–63, 217
Welansky, James, 132, 143, 161, 162
Weld, U.S. Atty. William, 208
Werner, William, 44
West Warwick, R.I., 174
Whitney, Josiah D., 28
Whitney, Louisa Goddard, 13, 16, 18, 19, 21, 23, 28, 218
William C. Hunneman Company, 32, 45, 46, 47. *See also* Hunneman fire engine
Wilson, Susan, 82n
Worcester Cold Storage and Warehouse Company, 216–17
Wooden chimneys, 4
Wooden rattles, 49
Woolley, Alderman William, 87–88

ACKNOWLEDGMENTS

This book would not have been possible without the expertise and generosity of Boston's firefighters and fire historians. My deepest gratitude goes to my upstairs neighbor Joseph Donnelly, a former Medford deputy fire chief, who opened doors for me into the fire-fighting community. Former firefighter Charles Kenney, an expert on the Cocoanut Grove fire, has been extraordinarily gracious in sharing his research and providing unstinting encouragement; Jack Deady, another Cocoanut Grove expert, has also provided invaluable background and information. Boston Fire Commissioner Paul Christian, an impressive fire historian himself, gave me much-needed guidance; I'll always remember his lecture on the meaning of the word *conflagration*. Former fire commissioner Leo Stapleton, author of numerous fire books, was an inspiration, as was former district chief John Vahey.

Special thanks go to Boston's Sparks Association and the volunteers at Boston's Fire Museum who donate time and energy to ensure that fire history is preserved. With deepest gratitude I thank Martin Sheridan and John Quinn, who shared their stories of surviving the Cocoanut Grove inferno. When Chelsea firefighters Herb Fothergill, William Coyne, and Leo Graves graciously talked with me about the 1973 fire, I could almost feel the air crackle with heat and sparks. Special thanks go to the extraordinary librarians at the Bostonian Society, the National Fire Protection Association, and the Massachusetts Historical Society. Among the firefighters, fire buffs, and other experts who deserve my thanks are Richard Powers, Ronald Caron, Patrick O'Rourke, James J. McCabe, Richard B. Magee, Bill Noonan, Elliot Belin, Gerard Crowley, Stephen E. McLaughlin, Wayne Miller, Larry Curran, Erik Anderson, Frank W. Fitzgerald Jr., John Esposito, Jay Hughes, Nancy Lusignan Schultz, and Edward R. Tufts. Special mention should go to J. Richard Hunneman Jr., Earl Doliber, Andy Swift, and all those trying to keep the old "masheens" running. Thanks also to Jane Alpert Bouvier, Kathy Alpert, and Nina Ricci of Gamewell. Special kudos go to Mary Prince for copyediting and cheerleading, to Renee DeKona for her excellent photography and to my editors and fellow scribes at the *Boston Herald* for their encouragement. Lastly, I must thank Commonwealth Editions publisher Webster Bull and managing editor Penny Stratton for their guidance and expertise, and Peter Cassidy for bringing this project to my attention.